Managing
Major Fires

By John F. "Skip" Coleman

PennWell Corporation
1421 South Sheridan
Tulsa, Oklahoma 74112
800-752-9764
sales@pennwell.com
www.pennwell-store.com
www.pennwell.com

ISDN: 0-912212-96-9

Cover design and book layout by Joey Zielazinski
Cover photo by Ron Jeffers

Printed in the United States of America
1 2 3 4 5 05 04 03 02 01

Dedication

This book is dedicated to the memory of my father. He retired the year I came on the job. Although I never had the honor of fighting a fire with him, I have been told time and time again that he was one hell of a firefighter and a good chief to work for. He was a good father and teacher. Chester, I don't believe I've ever told you, but "you done good!"

Table of Contents

Foreword ix

PART I
PREPARING FOR THE RESPONSE

Chapter 1
Review of the Incident Management System 3

Chapter 2
Sectoring Large Incidents 23

Chapter 3
Resource Allocation 33

Chapter 4
The Rules of Engagement –
Risk Assessment in the Fire Service 41

Chapter 5
Accountability at Major Incidents 47

Chapter 6
Managing the Mayday 59

PART II
THE FIREGROUND

Chapter 7
Fighting Fires in Older Apartment Buildings 67

Chapter 8
Fighting Fires in Garden Apartments 87

Chapter 9
Fighting Fires in Strip Malls....111

Chapter 10
Fighting Fires in Vacant Commercial Occupancies....129

Chapter 11
Fighting Fires in Institutional Occupancies....151

Chapter 12
Fighting Fires in Lumberyards....169

Chapter 13
Fighting Fires in Enclosed Malls....187

Chapter 14
Fighting Fires in Restaurants....205

Chapter 15
Fighting Fires in Churches....223

Chapter 16
Fighting Fires in Nursing Homes....239

Chapter 17
Fighting Industrial Fires....259

Chapter 18
Fighting Fires in Hotels....273

Chapter 19
Fighting Fires in Professional Buildings....291

Answers....305

Index....313

Foreword

It's been four years since I last sat at a computer, writing a Foreword to my first text. Since then, I have lost several more hairs. The ones I have managed to keep are opting for a slightly grayer look. Several inches have manifested around my mid-section and my siren-induced hearing loss is getting worse and worse–but enough about me!

Since the last book, almost five hundred brothers and sisters in our profession have lost their lives while working at what may be the most honorable profession on Earth. It's sadly ironic that this profession we all fight so hard to grasp and maintain still takes a drastic toll on those who love it so dearly. While I don't expect this text to eliminate our high death toll, I hope it may cut the numbers by making some of us stop and think before taking that next move.

This text centers on the fires that we all talk about by the kitchen table at the station during that golden hour after roll call, but before the housework starts. This is a text that deals with less than 1% of our working experience. However, these fires that I will be discussing are the reason that most of us joined this profession to begin with. This text deals with the larger, more unusual second and third alarm or mutual aid fires that happen less and less. The basics are discussed in this book along with some common-sense approaches to larger, non-routine fires. Most of us can put a "room-and-contents" fire in a single family dwelling out in our sleep. We do, however, respond to fires in non-routine occupancies (other than houses) on occasion. Today, these fires along with the routine house fires come at an ever-decreasing rate. Our chances to "play" and learn at these fires are becoming less and less frequent. Thc "old-timers" we used to soak up experiences from like a sponge are gone. Many of us are now those "old-timers".

I have attempted to do several things in this text. First, I have presented several chapters in the beginning of the book that discuss the organization and structure that should be present within any department in order to handle the bigger as well as the smaller, more mundane fires that we respond to. Incident Management is discussed as well as resource allocation. After that, I move into more safety issues such as Accountability Systems, The Rules of Engagement or Risk Assessment, and a chapter on Managing the Mayday. After this brief look at IMS and better "rules of engagement", I then concentrate on fires in distinct occupancy types.

Seldom do we sit at the table in the station and discuss the work we had in a Class-3 construction-type building last week. Instead we sit and talk about the vacant warehouse fire or the nursing home fire, etc. The last thirteen chapters deal with fighting fires in specific occupancy types. Construction concerns as well as strategic and tactical evolutions are discussed, and finally there is a case study at the end of every chapter. The case study is a fictional fire that is looked at from at least two, and often three, different views. First is the view from a small fire department perspective; about a dozen firefighters respond on this scene. Next is the medium-sized department with more staffing and firepower. In many examples, I also discuss big department operations with three engines, two trucks, and so on. I believe that this text offers different views of the same fire and hence provides a better understanding of the constraints faced by all three department types.

Safety, basics, and common sense are stressed throughout. There is always more than one right way to fight a fire. If we are to reduce the loss of life to brothers and sisters in our occupation, I believe we must look differently at the way we fight fires. Except in extreme instances, we must learn to fight fires on our terms and not the fire's terms. It is said that sometimes you have to slow down to hurry up! I believe that at fires sometimes we must slow down to all go home in one piece. This is not tantamount to cowardice. Rather it is learning to place additional honor in what we die and get crippled for. Taking great risk to save life is noble; taking the same risk for life or property already lost to fire simply is not.

Acknowledgements

There are many people I would like to thank for their individual efforts in helping with this book. As usual, I will surely forget someone and to that person, I apologize and offer my thanks, too.

First, to Theresa; thanks for giving me the time to work on this book. Thank you for your love and support.

To my children, Betsy and Toby; thanks for keeping it down and letting me use my computer every once in awhile.

To Joe and Mike; thanks for still keeping me out of trouble and keeping me sane. We all went from "A" shift to staff, but the friendship remains the same.

To Jim Murtagh, FDNY retired; thanks for letting me pick your brain and for working on a few scenarios with me.

To Ron Jeffers; thanks for the photos. They helped round out the book.

I would like to acknowledge Digital Combustion as the computer fire simulator used in the simulations in the text. They can be located at www.digitalcombustion.com.

Lastly, to about 500 of the men and women of the Toledo Department of Fire and Rescue (you know who you are); thanks for continually doing your best. You look good out there, which, in turn, makes me look good. It's an honor working with you!

Part One

The Form of the Response

Chapter One

Review of the Incident Management System

There are several truisms in the fire service. One that can be documented is that between 90 and 95 percent of the fires to which we respond are minor, if there's any fire at all. That's the good news. The bad news is that 5 to 10 percent of the fires to which we respond are all-hands or working fires. Statistically, about 0.25 percent of these working fires turn into major incidents. Whether your department responds to two or two thousand working fires a year, the clock is ticking. Your big one is just around the corner, and you'd better prepare for it now rather than during the incident.

The express purpose of incident management is to standardize the way of operating at every incident to which a fire department responds. The system outlines the roles and responsibilities of every unit that answers the alarm. One of its prime advantages is that it readily adapts to any type of incident, large or small, from an accident involving a child on a bicycle to the derailment of a train with hundreds of people on board. Incident management fits anywhere you use it at an emergency scene.

In my previous book, *Incident Management for the Street-Smart Fire Officer*, I presented the basic concepts of the system. I described it in the context of our bread-and-butter fires. In this text, I hope to expand on the system and present it in the context of both large and nontraditional fires. In teaching across the country and at the National Fire Academy, I am exposed to a wide range of thought on our profession. I find that most departments use an incident management system, but that many of them only use it at the bigger incidents, as well as nontraditional alarms. I am constantly amazed by the diversity of systems in use today. At this time, I do not see a standard form of IMS in the cards. There are too many personalities, procedures, and traditions in the way.

Such diversity isn't altogether a bad thing. What works in New York works there because that's the way they fight fires in New York. Likewise for Chicago, Phoenix, Toledo, and anywhere else you care to mention. There's always more than one right way to fight a fire. We may use different equipment, different evolutions, and different protocols, but the end result is invariably the same; the fire goes out.

No matter where you live, the potential for a large-scale incident is always present. The big alarms aren't confined just to the major population centers. Several years ago, I taught at the NFA with Chief Jim Conner, the first battalion chief to arrive at the bombing of the federal building in Oklahoma City. He told me over and over again that he had never expected that an event of such magnitude would ever occur in his town. Sound familiar? We should all realize that the next headline-maker might break in our own jurisdiction, and it could happen at any time. Natural disasters, fires, terrorist attacks, and airplane crashes can occur in any community, day or night. To remain ready for them, it's vital that we practice incident management, not only at the large incidents, but for the everyday calls as well.

Command Posts come in all shapes and sizes ranging in complexity from dedicated vehicles to simply a place to stand in the street. *Credit Ron Jeffers*

It *is* amazing how differently we fight fires across the United States. Some departments go in for aggressive strategies and tactics, while others exercise more restraint, more or less waiting until the fire comes to them. Some departments allow for different assignments at every incident. I refer to these as *incident-specific operations.* The first engine at the scene of one fire may be assigned to protect exposures, and the first engine at the next fire may be assigned to perform rescue, and no one ever pulls any hose off the rig. Some departments proceduralize their assignments. I refer to these as *pre-incident assignments.* The crew of the first engine on the scene normally goes to the fire floor for fire attack, while the crew of the second engine goes to the floor above the fire to check for extension. The first truck crew forces entry and then searches or goes to the roof, while the crew of the second truck ladders the rear of the building or assists in search. Some departments initiate the incident management system the moment an officer arrives at the scene. Other departments wait for the first chief officer to arrive, while still others wait for additional alarms. Some depart-

At this 3rd Alarm in Clifton N.J., the first-in officer assumes Command. *Credit Ron Jeffers*

ments use rapid intervention teams, while other departments can barely spell the term. Some use personnel accountability reports (PARs), while to others, the letters P-A-R mean a decent golf score.

Personally, I'm an eternal optimist. I know that no one individual will change the way that fires are fought across the country. Still, I believe that if enough of the key players in the various departments are exposed to different ideas, then some of the better ideas will begin to rub off. And one thing that I'm certain about is that none of us can afford to go about our business in isolation from the rest. The events that gain nationwide attention prove that.

The beauty of standardization is that, under incident management, everyone on the scene understands who is in charge, and everyone should have a keen understanding as to what his specific role is at a given incident. The system defines the roles. The first unit on the scene should establish some form of command and direct incoming crews to operate as dictated by the situation. No matter their function, everyone has a role and a subsequent responsibility.

Incident management provides checks and balances at every incident. This system of checks and balances is set in motion by the incident commander as he makes assignments, tracks the status of assigned crews, and evaluates the stream of information that he receives. In delegating responsibility, a typical commander uses command boards and flow charts to maintain a reasonable span of control, manage assignments, and pinpoint areas of weakness in the overall scene. It might be argued that this system of checks and balances could exist without a formal incident management system, and perhaps this is true. Still, it is within the framework of incident command that we can better guarantee consistency in operation and accountability in action, and it is this benefit that gives the edge to IMS.

A Command board is used to track crews at this working fire. *Credit Ron Jeffers*

One of the most common complaints that I hear about incident management is that departments don't always have enough people to fill all the boxes. Well, at times, neither do we! At times, neither does FDNY! Some people have this notion that if you use IMS, the first thing that you have to do at every incident is set up some grandiose command structure. By this philosophy, if an engine pulls up at a warehouse fire, the officer assumes the role of incident commander, the driver becomes the operations officer, and the third man takes charge of logistics. When the second engine arrives, the next officer becomes the liaison officer, the driver becomes the finance officer, and on and on. Finally, the third engine arrives on the scene, and its members are ordered to attack the fire.

What should actually happen is that the officer of the first-arriving engine should assume command and assign the remainder of his crew to securing the water supply. Their task is to connect to the sprinkler system and supplement it. The incident commander should then assign the remainder of the first-alarm units to attack, ventilation, and search. If a second alarm is warranted, the IC should ask the next-due chief officer to report to the command post and become the operations officer, and other incoming units should be assigned as befits the demands of the incident.

In such a scenario, the command structure was composed of two individuals (see flow chart on following page). All of the stuff identified in the typical IMS flow chart was fulfilled.

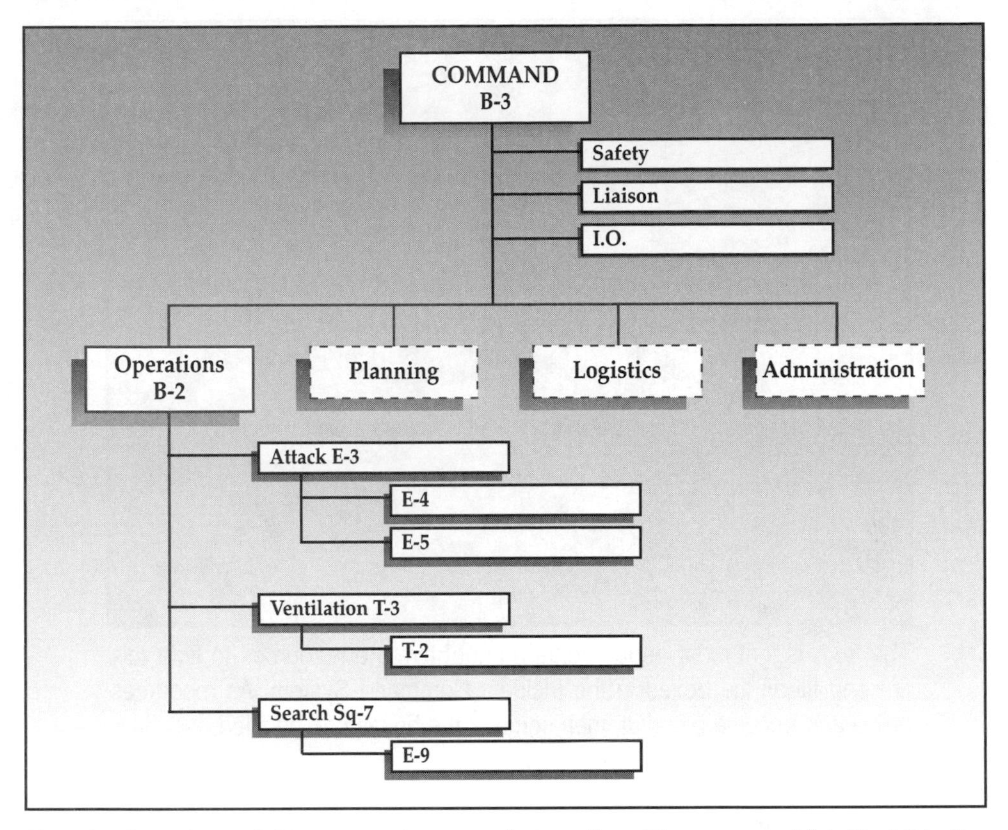

How the flow chart might appear at a second alarm fire. The command "staff" at this fire consists only of "command" and "operations". All the remaining "boxes" are handled by command.

That's the beauty of IMS: It's expandable! It's also unconstrained by time. There is no rule that says Box A must be filled within X number of minutes, or that Box B ever needs to be filled by an individual. I have been the incident commander at dozens of multiple-alarm fires, yet I have never assigned the role of administration officer. I, as the incident commander, have always handled that function. Now, in the western part of the U.S., at a large wildland fire, the role of administration officer may be the second assignment to be filled. In such cases, the IC needs someone to process the wildland firefighters responding to the incident. The system can be adapted to meet the needs of each particular emergency.

For incident management to work, it must be based on a written plan. Standard operating procedures relevant to IMS should dictate several key items, including when the role of incident commander is to be established. NFPA 1561 states that command shall be established by the first-arriving officer. As we all know, however, NFPA standards are just that—standards. They are neither law nor requirements. Still, it would be hard to defend in court why your department did-

The truth is that most departments do not have the resources to fight this fire and fill all the boxes in the Incident Command System. As resources arrive and become plentiful, then some of the boxes can be filled.

n't meet the minimum acceptable standards established by a group of peers in your industry. Thus, your department needs an SOP that states when command will be established at incidents. In Toledo, and in more and more departments across the country, command is established on the arrival of the first firefighter at the scene, regardless of rank.

Your SOPs also need to stipulate at what types of incidents your department will use IMS. Federal law (SARA III) mandates that such a system be used at all haz-mat incidents. The NFPA has set criteria for the use of IMS at structure fires. My department uses it for all emergency responses, regardless of the nature of the incident. Your department formally needs to establish its protocols regarding this matter.

At issue also is how to transfer command from one individual to another. At most incidents to which we respond, company officers arrive first. Several minutes later, a chief officer of some rank arrives. There needs to be a smooth transition in the transfer of power. Egos and rivalries need not enter into this equation. Transferring command merely for the sake of rank or ego is a mistake in that it often results in a missed opportunity for training and usually is a waste of time and effort.

The SOPs need to address important issues, such as staging and the method of dividing up an incident into manageable units, or sectorization. A few basic rules are in order,

As the incident builds, so should the "think-tank" at the Command Post. *Credit Ron Jeffers*

too. For example, no one should do anything until the IC tells him to. Does this mean that freelancing of any nature cannot be tolerated? Exactly. Does this mean that crews need to keep the IC informed as to what they're seeing and doing, as well as unanticipated changes in their actions? Absolutely. Does this mean that a company officer should abstain from informing the IC of a life-threatening hazard in his sector? No! The key to an effective organizational structure is in the flow of information to the command post. An incident commander cannot make reasonable decisions if critical facts about a given circumstance are withheld from him. Does this mean that sector officers need to inform Command of every little action and evolution that they're performing? Of course not! An incident commander can't afford to be swamped with minute details. That's why we train our officers.

It's a given that the incident commander shall be held responsible for the outcome of the incident. He is, in effect, the captain of a ship. If he's not responsible, then who is? Similarly, the IC is responsible for all that he does not delegate. If he doesn't assign an operations officer, then he must oversee those duties. The flow of information available to him should not be impeded, but he also shouldn't have to talk to anyone who isn't in his chain of command. If reports are coming in from the Operations sector, then the operations officer should handle them. In fact, it works best if the IC is on a separate tactical channel altogether. This helps to remove the temptation of micromanaging, or getting involved at the tactical level of an incident. The incident commander should remain at the command post. Many incidents are small enough that the command post can essentially be where the action is. Larger incidents call for the establishment of a formal command post, and the IC should stay there. As the hub of information and control, the command post is the most logical place for the IC to be, and it's the most logical place for an incoming officer to find whoever's in charge. I realize that this is in conflict with the practices in some of the larger cities in the country, but for departments about the size of my own and those smaller, I believe it works best if the IC stays put.

Under incident command, assignments need to be specific, clearly stated, and known to all responders, whether made prior to the response or after the arrival of the first unit. A doctrine of accountability must always be in force. The incident commander needs to keep up to date as to the status of all crews working at the scene. This is something that you can test. Go to your next working incident and ask the IC exactly where every crew is operating. Then, by observation, radio, or other means, check the accuracy of his answer. Is the accountability system adequate? Would you feel comfortable inside knowing how they're tracking you from the outside?

A chief officer and his aid discuss the fire. With the IC in a "permanent" location, company officers can always locate the person in charge. *Credit Ron Jeffers*

If you've heard me lecture or have read my other book, then you know my opinion on the importance of focus. I believe that IMS is built on the principle of focus. The task of a commander at any incident is to focus on the needs of that incident.

Such focus is best gained in the command post, where the IC can form a complete picture of the entire operation. Proper focus requires concentration and a direct thought process so as to anticipate the needs of the incident and translate those needs into the actions of sector officers.

I have heard of departments that advocate a lone-ranger type of incident command. They prefer that the IC be isolated. They reason that, with little or no stimuli from the outside, the incident commander can concentrate on the larger-scale aspects of the situation. Others believe that the incident commander can best serve the incident if he is directly in the line of fire, so to speak. Sorties into the hot zone and jaunts around the structure are believed to afford him a bird's-eye view of the scene, by which he can then make strategic and even tactical decisions.

What would some of the needs of this fire be?
Credit Ron Jeffers

Both modes are valid, depending on the circumstances. At large incidents, the IC needs to surround himself with a team, consisting of individuals whom he can trust and with whom he can work. This doesn't mean that only members from the same department can be on the team. In rural departments, the ability to assemble a team incorporating outside elements is vital. Chief officers of rural departments must develop trust and a good working atmosphere if incident management is going to work for them. True, the vast majority of working incidents in any community can be handled solely by the local department and without any help. As stated at the outset, this text is concerned with larger emergencies, and most of you can hopefully manage a single-family structure fire without breaking a sweat. When that rare but ever-looming large event does occur, however, you need to have relationships with outside agencies and other departments already in place. Build your team now, not on game day.

Even the best IC is only human, and no one can do it all. Great coaches know how to build teams and lead them to victory. Most realize, consciously or unconsciously, that focus is a matter of viewing a situation objectively, from the outside, not from within. Standing on the sidelines, allowing your other coaches and coordinators to do their particular job, allows you to see the overall picture. Don't micromanage! If your subordinates ask for input or advice, give it to them. If you don't know the answer, call a time-out and have a conference. For the most part, though, you should just give them room to work. You assigned each one of them a task. Now let them fulfill that assignment and make you proud.

Who should be the incident commander? In my opinion, the title should be established when the first officer arrives. That's for all incidents. As the operation grows, command should be transferred up through the hierarchy. At large incidents, there is a level at which it needs to stop being passed. For some departments, common sense and procedure indicate where that level is. For smaller departments, the incident will probably fall to the chief of the department.

For departments with only one chief officer, that's where it should fall. Build a team to fill the positions under you, utilizing your officers and those of mutual-aid departments.

For larger departments, having more chief officers, there can be some kind of ceiling at the upper level of command and a selection process that takes place there. In Toledo, it's rare for the director of the department to take command at an incident. Usually I, as deputy chief of operations, will arrive and take command from the line battalion that was originally dispatched to the incident. I then normally assign the Operations sector to that line battalion chief. Now I have a chief officer handling the line crews sent in to fight the fire, and I am responsible for the rest. There are several reasons why I take command and why I suggest that staff or higher-ranking chief officers of middle-to-large departments assume command at larger incidents. Let's face it, battalion chiefs are more accustomed to placing and working with companies. Coming from staff, I believe that I am more familiar with outside agencies and personnel. As a deputy chief, others expect me to be best equipped to handle the nuances of planning, logistics, and administration. If you're wondering why the director of the

The Phoenix Fire Department utilizes "Command Teams" at their large fires. This group of chiefs and company officers operate in the Command Vehicle. *Credit Steve Kreis*

department doesn't normally take command, I should say that I've never asked him, but I have noticed two things: first, that he has faith in his deputies to run incidents, and second, that he prefers to stand back and observe. He's there to support us, and if he doesn't like something he sees, he'll let us know. Normally, he just watches and listens. If you trust your people, have confidence in their abilities, and put them where they belong, then there's really no need to siphon off their authority.

As far as the selection process is concerned, keep in mind what type of incident it is and what chief officers are available. Incident types can vary widely, from fires to floods to vehicle mishaps to ice storms. Most chief officers have a particular area of expertise. At most large incidents, you have two key positions to fill: incident commander and operations officer, or sometimes planning officer. If I'm on the scene of a haz-mat incident and have a battalion chief also on the scene who is a haz-mat technician, you can bet that I'd use my authority to assign that chief to head up the Operation sector. At an airplane crash, I'd do best to assign a crash fire rescue specialist to the Operations slot.

For most of the incidents that we will discuss in this text, a formal command post will be used. A formal command post has several key components, the first of them being an incident commander. As stated earlier, I discourage the practice of moving around for anything other than the rare occasion. If an incident commander wants to know what's going on at the rear of an involved structure, he should get a report from an officer assigned there. If you really must look for yourself, transfer command to another chief and take your walk, but leave the source of power in the command post. Especially at large incidents, a lot of people want to talk to the IC. They shouldn't have to play hide-and-seek with whoever's in charge.

A formal command post also tends to be immobile. Lack of prior planning or a change in the wind may necessitate moving the command post. Otherwise, you should pick a location suited to the weather and the scope of the operation: big incident = big command post; little incident = little command post.

A formal command post is a place to think, write, and talk. Some departments can afford to have a command-post vehicle. Most of these have built-in communications, lighting, tables, chairs, and even bathrooms. The majority of departments can't afford such a luxury, however. Sometimes a street intersection, parking lot, or hotel lobby will have to suffice. At the past few large incidents that I've been to, I've used yellow scene tape to mark off the boundaries for the command post. It's amazing how much more controlled the command environment is, even with just a bit of tape strung along the perimeter. I haven't yet had to have a police officer maintain the entranceway, but I wouldn't hesitate to use one if it were needed. Try marking off boundaries with scene tape at your next large incident and see whether you don't gain an additional measure of control over the activities and personnel at the command post.

Within incident management, the accepted span of control is considered to be five subordinates, although this is certainly not a hard-and-fast rule. There are times when a span of

The Phoenix Fire Department utilizes "Command Teams" at their large fires. This group of chiefs and company officers operate in the Command Vehicle. *Credit Steve Kreis*

control of less than five will enhance operations, and there are other times when a supervisor can easily manage more than five subordinates.

Similarly, the span of communications is the number of individuals with whom one person can effectively communicate at one time. One of the key factors in this is a matter of atmosphere, or where the discussion is taking place. Are you in a factory with blast furnaces and punch presses pounding in the background, or are you in a quiet parking lot at four o'clock in the morning? There is an inverse relationship between the amount of distracting noise that your brain must filter out and the amount of information that it can process. The content of the message is also important. If the message centers on last night's bowling scores, then many can join in the discussion. However, if you're concerned with how much more stress a fire-weakened girder can support, or perhaps what's going to happen when spilled sulfur from a derailed railroad tanker mixes with diesel fuel and water, then only a few knowledgeable parties should be allowed to participate. When it comes to critical issues, be highly selective of your sources of information. Similarly, there are times in conversations when we must carefully choose our words so that those with less expertise can get the gist of what we're trying to say. Discussing options at a large incident with your peers requires less conscious effort than recounting the same topics to the media or explaining the fire and the tactics involved to a recruit firefighter. Finally, the mode of communication that you choose must be appropriate to the message. Alerting all on-scene crews to an imminent collapse isn't a one-on-one communication process. Conversely, inquiries about evacuation options should not be held over the radio.

You must consider the span of communication, especially at large incidents. Within the command hierarchy, the IC must be the conductor who leads discussions, determining the pitch and cadence of the dialogue. If things don't seem to be going smoothly at the command post, step back and look at who's talking and what they're trying to say. You might find that too many people are vying for the same spotlight. My instincts and experience tell me that, for the IC, the effective span of communications is about three to five individuals. For the operations officer, it's about the same number. For a sector officer in charge of a search in a high-rise, it may be only one or two searchers.

The IC's general staff consists of four entities: the four people who head up the Operations, Planning, Logistics, and Administration sectors. At large incidents, especially fires, the head of Operations is the role that I suggest be assigned first. The IC will have his hands full with a variety of tasks, and he'll need a ranking officer whose focus is on the fire. This will afford the IC the opportunity to stand back and focus on the incident as a whole.

In effect, the operations officer is responsible for the core of the incident (see the following flow chart). If the problem is a fire, then the operations officer is responsible for assigning crews necessary to put out the fire. If the incident is a train derailment with many injuries but no ensuing fire or haz-mat release, he is responsible for overseeing the EMS response;

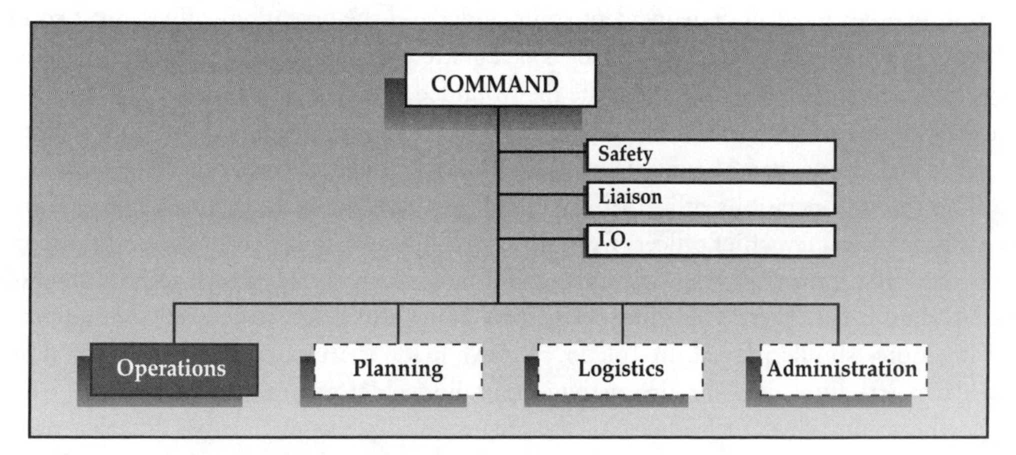

Flow chart highlighting the Operations section.

A Command Team works a fire from the Command Post. *Credit Steve Kreis*

specifically, the triage, treatment, and transport of injured individuals. This leaves the incident commander free to address other aspects of the emergency, which can be considerable.

Normally, the operations officer has the greatest number of crews and individuals reporting to him at an incident. The incident commander and the operations officer must continually assess whether the span of control is being stretched too much. Even with dozens of crews assigned to fight the fire or handle the wounded, the span of control of the operations officer shouldn't exceed around five subordinates. As we go on, I will illustrate how to ensure that the Operations span of control and communications isn't violated.

The role of operations officer can be filled by a number of individuals. Some departments have only a few chief officers from which to choose. Others have many. Some departments can fill all the slots from within and still have firefighters left over to hold the hoselines. Others don't. There is no rule, nor is there a suggested rank structure. Common sense and ingenuity should prevail. In Toledo, we have made it a matter of procedure that, at a two-alarm structure fire, the first-arriving battalion chief assumes command on arrival. When the second-due chief officer (or the Deputy of Operations) arrives, he takes over as incident commander and the first-due chief officer becomes the operations officer. The reason for this is that the first-due chief has already assigned crews and, hopefully, has a plan in place. To save time and explanations, the second-due chief becomes the IC so as to focus on the Planning, Logistics, and Administration sections, while the first-due chief maintains his original focus of attacking the fire.

In rural areas, such a protocol may not prove beneficial. The chief of a small department will probably use mutual-aid at a large fire. When the mutual-aid chief arrives, it may be more prudent for the home chief to retain command and assign the mutual-aid chief to the Operations sector. This makes sense, since the home chief is likely more familiar with the planning, logistical, and administrative aspects of his jurisdiction. With a little explanation, the mutual-aid chief can come in and concentrate on the fire.

The operations officer works at the command post with the incident commander. These two are joined at the hip. I know of some departments that don't believe it to be a good idea for these two entities to be working so closely together, especially at a high-rise fire. My concern on this issue is with radio traffic. It must be agreed by all that it is of absolute importance that the IC and the operations officer converse. Unless you have unlimited radio channels and people to monitor them, then it's a waste of a good channel for these key players to try to carry on an intelligent dialogue over the air. Build a good team and work with the members of it, not away from them. Entities that are assigned to the Operations section, including the staging officer, should communicate only with their immediate sector officer or the operations officer. The incident commander need not reply! The IC need not direct! The IC needs to stay off the Operations channel and keep focused on the bigger picture!

Consider the Planning section to be the think tank of the incident (see the following flow chart). The planning officer is responsible for planning the needs of the operation. This begins with a review of what has already happened. He should then assess what is currently happen-

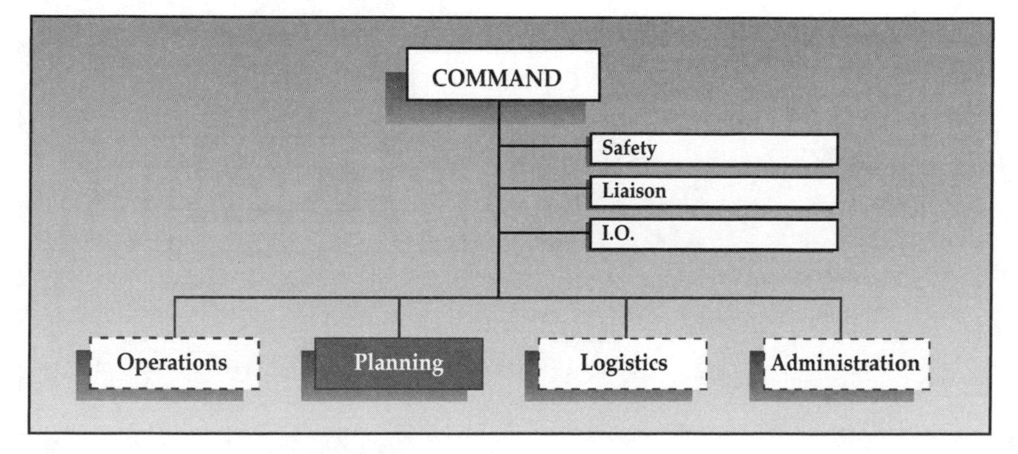

Flow chart highlighting the Planning section.

ing and what else can happen. Whereas the operations officer basically has one point of focus, the fire, the planning officer has several areas of concern. He, perhaps by means of a Resource Status Unit, must maintain an account of the resources used at the incident. This might entail establishing an Accountability section. Items that need to be accounted for range from personnel to end loaders to generators to foam supplies to lumber for shoring.

A Situation Status Unit keeps watch over the emergency itself, monitoring where the incident is going and what options are available. The Documentation Unit maintains complete and accurate records of the incident for legal and historical purposes, as well as for training, and the Demobilization Unit determines the method of de-escalation and dismissal as the incident draws to a close. Finally, any technical specialists also work within the Planning section. Such specialists can include environment experts, haz-mat experts, construction experts, legal experts, and water supply experts, if there are concerns about the water supply. (Under normal circumstances, when the water supply doesn't pose a particular problem, the water supply officer should work within the Operations section.) Although I have never seen it in writing, I believe any investigation into the cause should fall under the Planning Section.

The logistics officer is responsible for getting the necessary tools, equipment, and qualified operators to the scene (see the following flow chart). Because there are six units under Logistics, this section can be divided into two branches. When used to the fullest, each branch is under the direction of one branch director, who should report directly to the logistics officer. The first of these branches is the Service Branch.

The units incorporated under this branch are the Communications Unit, the Food Unit, and the Medical Unit, including Rehabilitation. The second branch is the Support Branch, which covers the Supply Unit, the Facilities Unit, and the Ground Support Unit. Briefly, the Supply Unit is responsible for ordering non fire-service personnel, supplies, and equipment;

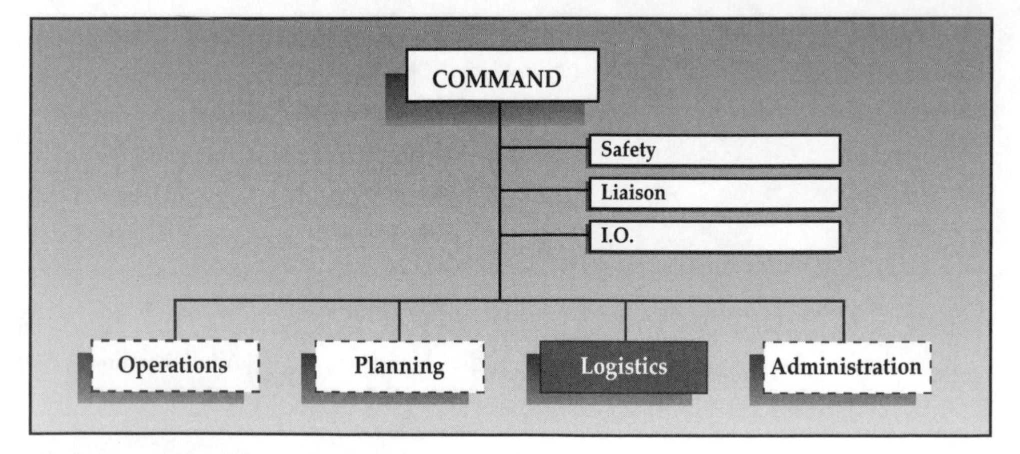

Flow chart highlighting the Logistics section.

the Facilities Unit is charged with scouting out suitable buildings to house the command post, field hospital, morgue, rehab center, and so on; and the Ground Support Unit handles vehicle maintenance and transportation at the scene.

I have been a chief officer for over fourteen years now, and I've responded to hundreds of working fires as a chief. Never once, in all of those instances, except in drills, have I ever used a Logistics section. Still, I'm sure that I will, sometime before my career is over. A logistics officer brings something to those one-in-a-million incidents that can spell the difference between success and a mess, and it's something that isn't even mentioned in the books. That something is *time.* A good logistics officer can save an incident commander time in figuring out how to get fans and cranes and end loaders and the people to run them. A good logistics officer can even make a difficult incident seem better, simply by providing a place for your crews to go and relax.

I don't believe that I will ever set up an Administration section, however. As I mentioned earlier, out West, where they have incredible wildland fires, departments set up an Administration section almost as soon as they establish command. They need an administrator to hire personnel and track work hours for the hundreds of wildland firefighters who show up to fight the fire.

In metropolitan areas, the Administration section isn't something that gets established often, if at all. Most of our incidents don't last for days. We fight fires with the crews that we have on the current payroll, and we don't normally hire firefighters on the spot. Not normally!

In its full incarnation, the Administration section controls four units (see the following flow chart). The Time Unit records the time cards for payment of personnel. The Procurement Unit handles purchases from and contracts with outside vendors. The Claims

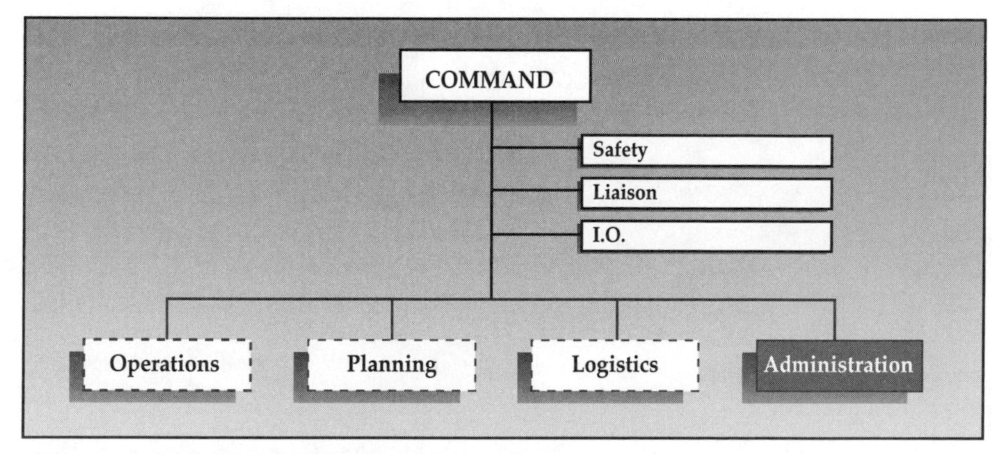

Flow chart highlighting the Administration section.

Unit takes care of workers' comp claims for injured personnel and may operate in conjunction with civilian insurance companies. Predictably, the Cost Unit collects cost data so that the incident commander and the planning officer may better assess whether a given mode of operation would be worth the expense.

There are three other positions for which the incident commander is responsible (see the following flow chart). At normal fires and other incidents, he can handle these positions, along with the roles and responsibilities of his general staff. These positions aren't considered to be part of the IC's span of control. Those who fill them report directly to the IC and simply help to make the incident run more smoothly and effectively.

The first of these positions is that of safety officer. The IC is responsible for scene safety. However, when he cannot devote enough time to the safety aspects of the incident, he should designate a safety officer. My department has twenty-four-hour safety officers. One responds to every structure fire that we have. Safety is also dispatched on special types of calls, such as haz-mat, water rescue, and mass-casualty incidents. The safety officer should roam the scene, looking for problematic areas. Although he reports to and is generally outranked by the incident commander, the safety officer may order an operation stopped. The IC always has the final say, but he would be hard pressed to go against the concerns of his safety officer.

The second position to note is that of liaison officer. This entity is responsible for interacting with the outside agencies that report to the command post and either have a monetary interest in the incident or are called in to assist. The primary role of the liaison officer is to see that representatives from outside agencies find their place in the incident. By acting as a buffer and an information source, the liaison officer can also keep a lot of these outside reps from disrupting important conversations at the command post.

The third position is that of information officer, who has many tasks at a major incident. Along with the obvious task of providing the news media with updates, the information officer

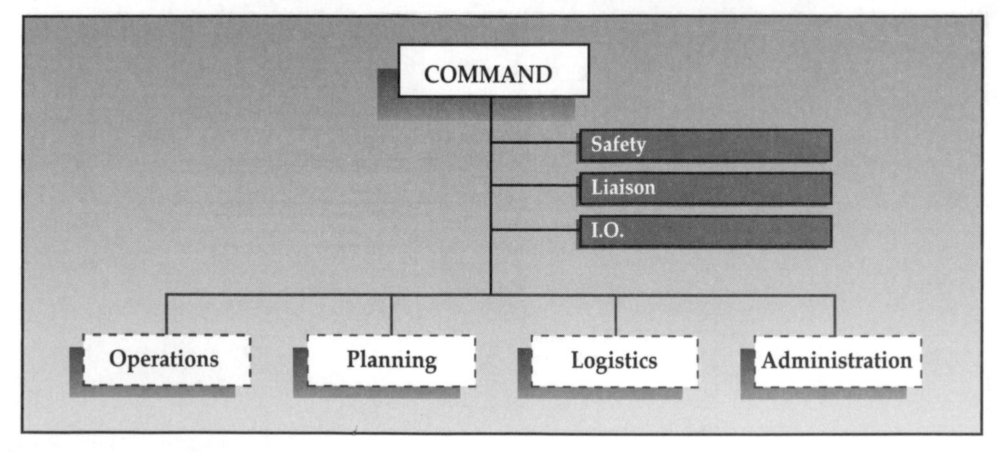

Command's staff.

Span of Communication relates to the number of individuals one person can effectively communicate with. *Credit Ron Jeffers*

is now also used as a contact with relatives at the scene, as well as civilians who are affected in one way or another by the incident. The information officer is also a contact for relief crews, briefing them on the incident and whatever progress has been made since the last operational period.

Chapter One Questions

1. The Incident Management System defines specific ____ of everyone responding at a given incident.
2. What is the specific NFPA standard that deals with the Incident Management System?
3. According to the author, the Incident Management System is built on the principle of _________ .
4. What is considered the normal span of control under IMS?
5. List the four Command General Staff positions.
   a)____________________
   b) ____________________
   c) ____________________
   d) ____________________
6. Operations is responsible for the _________ of the incident.
7. The author recommends that the Operations officer work at the _______________.
8. ________________ is responsible for getting the necessary tools, equipment and qualified operators to the scene.
9. Name the three Command Staff positions
   a) ____________
   b) ____________
   c) ____________
10. Which of Commands staff works with outside entities and helps them find their place in the incident?

Chapter Two

Sectoring Large Incidents

The word sector has been overused and misused throughout the country. If I were to offer a definition, I would define it as "a generic term that refers to a functional or geographic assignment." When we speak of a sector at an incident, we're referring to an activity or location. Even at that, a sector isn't bound to one specific place or action. A sector assignment may involve a myriad of tasks, many of which might be carried out at different points around the scene, all of them aimed at the same basic goal. Keep in mind that the word *sector* isn't equivalent to *task*. Efforts should be made to stop confusing the two, referring to tasks as the "vent sector," the "rear sector," and so on. There are better, more appropriate terms to describe an activity or a physical location—terms that are nationally accepted.

As per the National Fire Academy, there are two generally accepted types of sectors used at all incidents. I would like to add a third type, which is a combination of the others. Although not recognized by the NFA, the third type of sector that I'll describe simply helps define expectations, and isn't that the underlying purpose of a sector, anyway? to define expectations? If you're assigned a sector, regardless of type, you immediately know what's expected of you. Rather than freelance and do what you believe is best, you accept an assignment from a higher authority. This ensures a controlled, coordinated, balanced operation—one that proceeds from a plan.

The two types of sectors recognized by the NFA are functional sectors and geographic sectors. As can be deduced by the name, a functional sector is activity-oriented. Functional sectors, referred to as "groups," fall under the supervision of a group supervisor. Like the IC, the span of control for a group supervisor should be no more than five. This means that one group supervisor can have five companies or crews, each with five crewmembers, working directly under him. Even for a department the size of my own, that is a significant amount of staff.

Some examples of functional sectors include entry, search, rescue, ventilation, extrication, and recon. Unless otherwise specified by the incident commander or operations officer, functional sectors are responsible for their activity throughout the scene. This means the entire area, the entire structure, or, in the case of a traffic accident, all of the vehicles involved. If fire involves a twelve-story structure and the IC assigns a single crew to search operations, then that crew is responsible for conducting a search on every floor of the building. That would be a pretty tall order (no pun intended), and there is certainly a limit as to how much one crew can search. If the IC assigns a heavy squad to extrication, then that squad is responsible for extricating all of the victims in all of the vehicles involved in that accident.

A geographic sector is used when the incident commander wants a crew to work in a specific area. These sectors are normally referred to as divisions and fall under the direction of a division supervisor. When the IC assigns a geographic sector at a structure fire, the division supervisor is responsible for all of the activities required in that division or on that floor. A division supervisor assigned to Division 16, the sixteenth floor, is responsible for attack, overhaul, salvage, and all other operations there.

In a scenario such as this one, the division supervisor wouldn't be responsible for any functional assignment made on that floor. For example, if the incident commander were to

Crews operating in the front of this North Berger N.J. fire might be considered operating on "side A". Command may have ordered the building laddered on "side A". *Credit Ron Jeffers*

assign a ventilation group, then the division supervisor needn't conduct a vent operation in his area. However, his extinguishment efforts need to be coordinated with those of the vent group supervisor. In short, a division supervisor is responsible for all activities in his area except for those that have been assigned as functional sectors. Because of this, the IC must keep his division supervisors informed as to any additional teams that he sends into operation.

Normally, when the word "division" is accompanied by a number, it indicates the floor of the structure. If a letter of the alphabet accompanies the word division, it refers to a particular side on the exterior of the structure. "Roof" is fairly self-explanatory, except that, if the IC were to assign both a vent group and a roof division, I would assume that the former would be venting the structure (including the roof) by some means, and that the latter would be overhauling the roof area. At the derailment of a passenger train, a geographic sector might be established for each particular car, designated by the numbers on those cars.

One topic that is open for discussion has to do with exposures. Traditionally, these are neither referred to as a group nor as a division. Exposed buildings cover a specific area, and as such, might be considered a geographic sector. Crews assigned to protect exposures normally handle all of the needs of the buildings in question, and this too bears a relationship to

All crews operating in the rear of this fire should be under the direction of one officer who's radio designation could be "Side C". *Credit Ron Jeffers*

the concept of a geographic sector. Still, these assignments are normally referenced only by the word exposure, plus a letter designation that corresponds to the adjacent side of the involved structure.

We must make sure that all firefighters assigned to this fire are "singing off the same song-sheet". *Credit Ron Jeffers*

The third type of sector, in my opinion, is the combination sector. There are times, albeit few, when I want to fine-tune an activity within a particular area. Normally, search, rescue, and extrication ops are the ones that I will most commonly designate combination sectors; i.e., the conjunction of a functional and a geographic sector assigned to one or more crews under the direction of a division supervisor. I call them division supervisors because, when a combination sector is assigned, it's normally assigned by stating the geographic area first, followed by the activity to be performed there. An example of a combination sector would be Division 2 Search. Such a crew would be responsible for search on the second and only the second floor. A crew assigned to East Stairwell Rescue would be responsible for the removal of all located victims in that area. A crew given Train 19943 Extrication would be responsible for extricating all entrapped victims on that railroad car.

Combination sectors fine-tune expectations. If assigned, their existence must be made know to all division supervisors operating on the same division. If a group has been assigned to the incident, these members must also be informed that a particular crew will be handling their task on a specific floor or division.

You may wonder why I would even consider assigning a combination sector at an incident. The reason, in my opinion, is one of time. Suppose that you are the IC at a working fire in a six-story apartment building. Heavy smoke fills the third floor. The fire seems to be contained on the same level, and you get reports that there may be victims still on that floor, as well as on the fourth. You assign a crew to Division 3 and one to Division 4. Your expectation is that they'll fight the fire on the third floor and handle any extension on the fourth. You know that, given the size of the fire and the smoke conditions, if any search is going to

Combination sectors would fine-tune expectations in this fire in a six-story apartment building.

be successful, it must be done as soon as possible and by members who don't have a hoseline in their hands. You assign your next crew to Division 3 Search and the next available crew to Division 4 Search. With a dedicated team on both floors, any victims there will stand a better chance of being located and removed in time.

If a single crew were assigned to perform the search, the likelihood of finding savable victims might be diminished. If a single officer and his crew are assigned, then that officer has some very important decisions to make quickly. As he walks up to the structure as a group supervisor, his first consideration is for the safety of his crew. If he believes that he can enter and operate safely without the benefit of a hoseline, then his next concern is to assess the building and the specific area that needs to be searched. He must then relate what he sees to the resources required to complete the task. This might mean a single crew for the entire structure or several crews for each specific floor or area. Time is the deciding factor here. If your aim is to pull viable victims from the structure, then the area to be searched must be covered within the next few minutes, depending on conditions and the capabilities of the searchers.

Since sectoring provides a system for identifying areas and responsibilities, each sector should be specific and identifiable. Use common, standard terms known to all who respond. There will be times when off-the-cuff terminology is necessary. Sometimes we get into situations that we could never have anticipated. Plan for the common, and improvise for the rare. The key is to make sure that everyone on the scene understands the terms that you're using. Draw a diagram, if you must. When dead airtime is available, make a general announcement that "the north wing of the structure where we're taking the victims will be called the north wing." Your intentions can only be acted upon if you communicate them intelligibly.

What functional sector assignments should be made at this fire? *Credit Ron Jeffers*

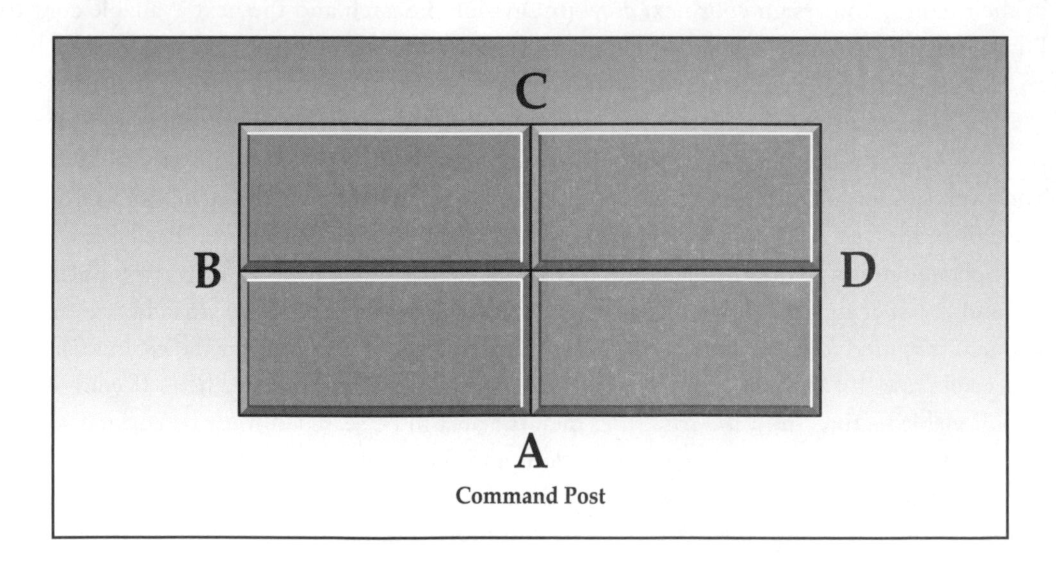

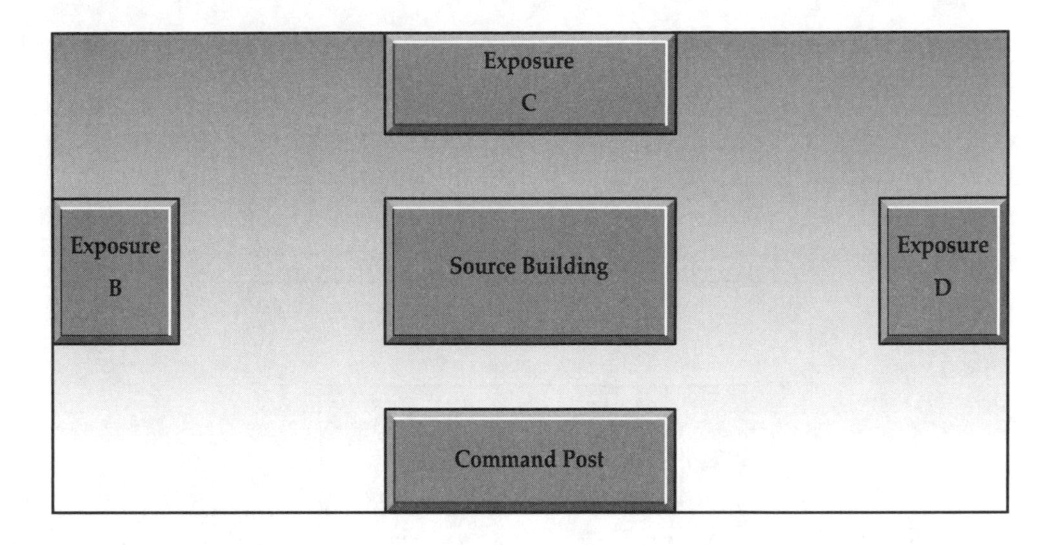
Exposure
C
Exposure
B
Source Building
Exposure
D
Command Post

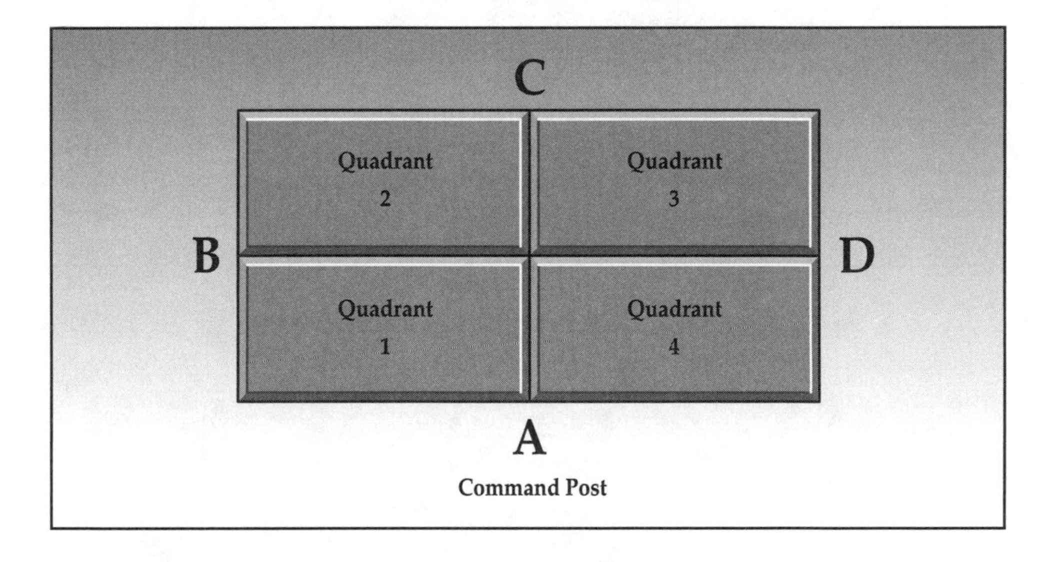
C
Quadrant
2
Quadrant
3
B
D
Quadrant
1
Quadrant
4
A
Command Post

Exposure B-2	Exposure B-1	Exposure B	Source Building	Exposure D	Exposure D-1	Exposure D-2

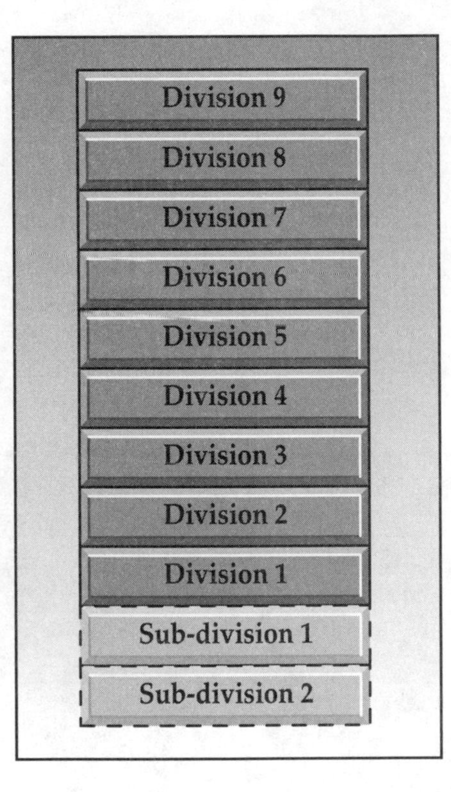

Chapter Two Questions

1. ____________ is a generic term that refers to a functional or geographic assignment.
2. What are the two accepted sectors used by the NFA?
   a) _______________
   b) _______________
3. Which of the above answers in question 2 defines an activity?
4. Which of the above answers in question 2 defines an area?
5. Why does the author also utilize Combination sectors?

Questions for discussion

a) Compare and contrast the differences between your departments use of sectors and what the author and NFA advocate.

b) Why does the author stress that all departments that respond must understand what other departments call "stuff" and how well does this work in your department?

Chapter Three

Resource Allocation

If you're the chief of a large department in one of the major metropolitan areas, the subject of resource allocation will have a different connotation than it would if you were the chief of a rural department with only an engine or two. One chief rarely receives mutual-aid but is likely asked on occasion to provide it, while the other gives and receives mutual-aid on a regular basis. Such departments are at opposite ends of the spectrum, yet both must consider resources and where they can get them.

This text deals with large incidents. A general rule in the fire service is to gear for the normal, not the extreme. Although we in Toledo, as anywhere, might potentially experience a major catastrophe, we only maintain what will be required to handle the "normal" emergencies that we experience in our city. At times, we're all sitting in our stations, waiting for the next run. Other times, all of our units are out on the street, in or out of service; committed to an incident or returning from the last one. The times when we're actually stripped of all of our resources is rare. Firefighter unions contend that we never have enough members to maintain appropriate staff levels for the safety of the members and the citizens. Management contends that, on average, resources are adequate for the need. Both are correct from where they stand. When called to those extreme situations, however, an incident commander must have an understanding as to what resources are required, what resources are available, and under what time frames he must operate.

Each type of incident requires certain resources. What's needed to extinguish a fire in a twenty-story office building may be useless at a major haz-mat incident. One requires many firefighters and a lot of water, whereas the other may require only a few qualified technicians and a nonsparking wrench. In both examples, I, as incident commander, want either the crews and the water or the wrench on the valve right now! In a big city, that may not be a big problem. In a small, rural department, that may be impossible.

The answer isn't to have all of the resources that a big-city department would have. There are times when even the chiefs of the metropolitan departments don't have enough. The real answer is twofold. You have to know what you need and where to get it.

What you need is probably specific to the incident. A good incident commander knows what he needs and realizes it early enough to get it on the scene in time for it to be effective. Knowing where to get two hundred gallons of foam is one thing. Knowing how long it will take to bring it in and how long it will take to apply it is quite another.

It would be a waste of time to create a list of incident types and some of the unusual items that you might need for mitigation. I could never think of them all, and neither could you. No one can. The best that you can do is surround yourself with a team of individuals who can help you identify the needs of any particular incident and who can help work miracles for that one-in-a-million call.

What would your estimation of the resource needs of this fire be? *Credit Ron Jeffers*

The average bedroom fire requires (a) one engine that can deliver at least 500 gpm, (b) a crew of four members, or two in and two out, (c) one 1 3/4-inch hoseline capable of flowing 48 gpm (12' X 12' = 144 / 3) and equipped with either a combination or a smooth-bore nozzle, and (d) one PPV fan. In Toledo, we would get a first-alarm assignment for such a fire with approximately twenty-three members responding; i.e., three engines, one truck, one heavy squad, a chief, a life squad, and a few other single-person units. For a room-and-contents fire, we would probably wind up sending some of these units home without using them. In a rural area, two engines could probably handle this fire, but those crewmembers would be a little busier than the guys in Toledo. When it comes to departments that can only muster three or four members, the fire can be held to one room, and it will go out, but those three or four firefighters are going to be very busy for a while. My point is that over ninety percent of the fires to which we respond can be handled with what we have.

A small section of a petroleum refinery in Toledo. Think of what could happen and then what would be needed. *Credit Coleman*

And what of the rest? To truly look as if you know what you're doing, know in advance, as much as possible, where to get the additional personnel and materiel you might need. This means not only where, but also how long. Two hundred gallons of foam will be useless if you have to wait all day to get it. Plan your potential resource needs. A fireboat won't be a required resource item in Los Alamos. Look at what's out there in your jurisdiction. Are there strip malls? Is there an enclosed mall? A refinery? A university with dormitories? Are you near a fault? Are tornadoes a possibility? If your needs have to do with staffing, consider mutual-aid agreements. I'm still trying to get mutual-aid agreements with some of the larger towns around Toledo. I'm talking about Detroit, Columbus, and Cleveland. You might say that it would literally take hours for any of them to get to Toledo if they were called. I would answer that it took sixteen hours to bring the 1 Meridian Plaza fire under control in Philadelphia, and they had hundreds of firefighters on duty. To me, two hours' travel time might be acceptable at the big event, in the grand scheme of things.

I should point out that the last I knew, Philadelphia doesn't have any mutual-aid agreements. Their philosophy is that if they can't handle it with what they have, then it can't be handled. What works in Philly works in Philly, I suppose, but most departments don't have their resources.

In Toledo, we currently have mutual-aid agreements with more than twenty departments in the area. As good as these departments may be and as close as they are, if Toledo ever experiences a disaster on the scale of the 1 Meridian Plaza fire or the Oklahoma City bombing, those twenty departments won't be enough. At an incident of that sort of magnitude, a response time of three or four hours from Columbus doesn't seem bad at all.

Staffing in Toledo is set at a minimum of 103 per day. That figure is based on a daily average call for service. There are times, approximately once a month, when this figure is too small. During multiple alarms, we may need more than that minimum. During thunderstorms with numerous lightning strikes and heavy wind that results in downed wires, the same holds true. There are other times when every unit in the city is available. When this happens, 103 is probably too many. New York City has over 1,200 firefighters on duty at a time, and sometimes 1,200 isn't enough for them! So look at the potential in your jurisdiction and ask yourself what you'll need for that worst-case scenario. After you've identified those needs, don't run out and requisition a whole bunch of stuff that you'll seldom use. Instead, figure out where you can get it in a pinch.

For example, look at the services that your town provides. Does your town have access to cranes and wrecking balls? Does your town have end loaders and backhoes? Come up with a

Some departments have ships that must be considered as part of their "total" fire problem. *Credit Coleman*

resource list of the tools and equipment available in your area, then a phone list of who to call at 3:00 a.m. on a Sunday morning to get it out to the scene. Take into account who'll operate it once it gets there. One rule of thumb for the fire service is that no firefighter should ever be allowed to operate a forklift. He'll probably do more damage than if you were to let the stuff in the warehouse burn. Make a list of operators, or let the foreman of the DPW arrange a pool of operators for you. Teach those operators how to use SCBA. They won't be doing searches with them; they'll be driving an end loader or a forklift, moving stuff for you. The thirty or forty minutes that it'll take to teach them how to use SCBA and perhaps change the bottle will be time well spent. If this isn't feasible, have a firefighter ride alongside the operator while he's working.

Radios are always a consideration at any mutual-aid call, and you must do all you can to ensure that your systems are compatible. Without common frequencies, extra portables may be a must. If your department is like mine, however, you may not have an abundance of extra portables. If this is the case, assign crews from your department to the mutual-aid companies, or at least a firefighter with a compatible portable.

When bringing outside firefighters to the scene, make sure that everyone speaks the same language. The protocols of IMS and sectorization will help in this regard. Normally, the

The skyline of a "not so big" city. Are the resources available to handle a fire in one of these high-rise buildings? *Credit Coleman*

neighboring departments with which you regularly respond will already be in tune with your terminology. This may not hold true for departments that you call in from afar to fight the big one. At best, meet with these other entities in advance and exchange cheat sheets. At worst, send a representative to staging and tell them what you want. Assign a firefighter to those crews to ensure accurate communication with them at all times.

Try to remove local colloquialisms from your day-to-day operations. For example, in Toledo, we call a short extension ladder a "dinky," whereas other departments tend to refer to these as attic ladders. Use functional assignments as much as possible. Instead of using unit designations, assign crews by task, such as Vent Group, Attack Group, and so on. Such designations are easier to understand. Most smaller departments use three-number designations to differentiate units, but it isn't readily apparent whether Unit 331 would be an engine, a truck, or a chief.

Recently I read an SOP from a major department. It discussed actions to be taken after the receipt of a 10-45 signal. After the 10-45, the SOP stated to wait for a 10-72 from the truck company. To me, it makes more sense to speak in plain English. Think about it! Civilians who buy scanners will surely learn in a short period of time what a 10-45 is, anyway. Realize, too, that brevity isn't ever really sacrificed with plain English. Translated, the code 10-45 has just three syllables to pronounce: "fatal fire." I realize that officers don't just say "fatal fire." They say, "We have a Fatal Fire." Conversely, crews that use 10-codes don't simply state the code. They also say, "We have a 10-45," so there isn't much to be gained by being cryptic. What could be lost is comprehension at a major incident. "Evacuate the building" will be understood by everyone at the scene, regardless of where they come from. A call for "All units, 10-76" may not be so readily understood.

Chapter Three Questions

1. Who is ultimately responsible for getting what is needed to an incident?
2. The author states that a general rule in the fire service is to gear for the _________ and not the _____________.
3. The key to understanding resource needs is to _______________
4. Can resource needs be affected by mutual-aid agreements?
5. What can preplanning do to your resource needs?

Questions for discussion

a) How can mutual-aid agreements enhance resource needs?
b) What role do pre plans play in resource needs?

Chapter Four

The Rules of Engagement – Risk Assessment in the Fire Service

Though brief, this may be the most important chapter in the book. This chapter will help define the "Rules of Engagement" for the fire service. At times, the hardest part of firefighting is in trying to figure out which type of ventilation to use, or which nozzle is best for flashover. At other times, you just don't know whether to press on with the offensive or go defensive. Many factors enter into these sorts of decisions. Ego, machismo, and gut feelings all play a tug-of-war with intelligence, common sense, and training. There are fires that go out prior to our arrival, and there are some that just refuse to go away until they die down on their own. And you, as the ranking responder, must figure out your course of action in a very public arena.

The National Fire Academy has a sample risk policy in its curriculum. That policy is as follows.

> We will take great risk to save life.
> We will take minimal risk to save property.
> We will risk nothing for lives or property already lost.

This policy speaks volumes about the modern realm of firefighting. These words may be easy to memorize and understand, but at times, they're very hard to follow. Leadership must be exhibited in those cases where life or property is already lost. This is where experience, firm decision making, and constraint must be exhibited. Those three sentences can define the rules of engagement for the fire service.

Let's look at this policy in depth. When I pull up to a fire, I look at the picture in front of me. The first thing I ask myself is whether we can enter and operate effectively inside. If the answer is no, then I have to assign activities directed at changing the conditions that prohibit

As it relates to a Risk Policy, "We will take great risk to save life". In this fire, great risk is warranted. The fire appears to be in a single family occupied ranch house.

interior operations; otherwise, we'll have to wait until those conditions change themselves. If the answer to the above question is yes, my next question is, "Are there savable people inside?" If the answer is no, I won't allow crew members to put themselves in danger searching for and removing deceased victims. Again, we'll either assign activities directed at changing the conditions inside, or we'll wait until those conditions change by themselves. If there are no savable victims, we will fight the fire on our terms either offensively or defensively.

When the picture in front of me indicates that there are savable persons inside, then I will allow crews to take greater than normal risks to locate and remove them. Our protective equipment gives us an edge over the fire. Depending on the circumstances and available crews, we may remove victims, then pull back to fight the fire on our terms. Other times, we may commit all of the on-scene crews to putting out the fire and ventilating, then go back later to rescue the victims. These decisions are up to the incident commander, based on his knowledge, experience, and gut feelings.

With no life hazard present, we should make every reasonable effort to save as much property as possible. Interior lines should initially be positioned so as to cut off the spread of fire by confining it to the smallest area possible. Crews in full, approved turnouts and equipped with SCBA can make interior offensive attacks on the fire. In fires where there is

In this fire the taking of "some risk to save property" would be warranted. This occupied Victorian home has heavy fire conditions on Division 2. Interior operations exercising caution and proper ventilation techniques would be warranted.

savable property to protect but the interior conditions warrant caution, you should place backup lines right behind the advancing nozzle. Assign rapid intervention teams and have them ready and standing by at the command post. This is what is meant by taking minimal risk to save property. Attack the enemy, but do so on your terms, not the fire's.

If a vacant building is totally involved on arrival, consider it to be already lost. Vacant buildings that have suffered fires in the past fit into this category also. Occupied buildings that are totally involved on all floors are also goners. We can only do and save so much. Even when the enemy is so strong that there is no chance to save life or property, we must still fight the fire on our terms. Vent the structure or wait until it vents itself. Cool the fuel or let it burn itself out. Set up lighting prior to entry. Knock down the fire from the outside, then go in to achieve final overhaul.

My goal each working day is simple: to protect as much life and property as possible, then go home. This applies to everyone with whom I started the shift. Walking up to a front door at three in the morning to tell a new bride that she is now a widow is never an easy task. Bearing the same message when the member died while trying to save property beyond repair is almost impossible. When a life has been expended in a battle over nothing of value, no explanation will suffice. Tragedies of this sort simply make no sense.

In this last example used to illustrate a Risk Policy, we have a vacant commercial building with heavy fire on the upper portions of the building. In this case, "nothing should be risked to save life or property already lost". Exterior operations with appropriate collapse zones should be established.

Risk assessment entails training officers to make educated, realistic decisions about expected outcomes. Through fire simulation, case studies, and other modes of training, officers of all ranks must come to make risk assessment a habitual process. If the mission at hand is worth the risk, then the operation should continue. If not, the officer should change the conditions or wait until the conditions change.

Risk assessment also means having a realistic understanding of the nature of fire protection in your community. This is reflected in response time and resource allocations. What is the predictable response time in your jurisdiction? Two minutes or ten? How long does it take your 911 operators to forward the call to your dispatchers, and how long does it take your dispatchers to pump out the run? Look at target hazards in your community and ask yourself whether you can protect all of the occupants there. If fire breaks out in the lower portion of an apartment building, how long will it take your department to search all twelve apartments on all three floors? If you don't have the resources to search effectively and stop the fire, what is the best course of action that you can take? Do you know?

Have a realistic expectation as to what crews can accomplish. We can't walk through fire for very long. We can only search for fifteen minutes if we use thirty-minute bottles. A two-person crew can't pull and advance a charged 2 1/2-inch attack line up the stairs of a tenement house. Can your crewmembers search that twelve-unit apartment building in fifteen minutes? Do time studies of your crews and make your decisions accordingly.

The art of risk assessment is in combining all of the parameters described above. Know your fire problems, your department, your officers, and their capabilities. Educate yourself through experience so as to form realistic expectations as to what crews can accomplish for you at your next fire. Depending on conditions, your crews can either successfully handle a situation or compound the problem. A good chief or company officer knows in advance the probable outcome and makes his assignments accordingly. There is no cowardliness in admitting that nothing can be done to reclaim property that has already been ravaged by flames. We can't save 'em all. And we shouldn't add to the toll exacted against our brethren merely for the sake of pride, ignorance, or both.

Chapter Four Questions

1. Write the NFAs' risk policy.
2. When the author discusses the risk policy, what is the first question he asks himself when he pulls up to a fire?
3. If a vacant building is totally involved on arrival, consider it to be ____________.
4. Risk assessment entails ______________.
5. According to the author, when a building is well involved upon arrival, Command must ______ conditions or wait till ____________.

Questions for discussion

a) The author states that Risk assessment involves educated realistic decisions about expected outcomes. Discuss initial decision making at a fire and compare those decisions with expected outcomes.
b) How can understanding what crews can accomplish help determine risk assessment?
c) Should staffing have anything to do with risk assessment? If you had twice as many members responding, would your decisions be different or are we talking apples and oranges?

Chapter Five

Accountability at Major Fire Incidents

Both NFPA Standards 1500 and 1561 address accountability systems for all incidents. Accountability systems are relatively new manifestations of the firefighter safety revolution of the 1990s. They will continue to be refined and standardized long after this text has become outdated. They are proving to be vital safety tools, despite some drawbacks inherent to the systems in use today.

In some respects, these systems are very personal items. When it comes to firefighting gadgets, some departments have seemingly unlimited funds for whatever is in vogue. Other departments take a more modest and fiscally conservative approach toward innovation, waiting until each new tool or procedure has been tested by time and trial. Some accountability systems can cost thousands of dollars to get them up and running. Currently several companies are selling GPS systems to track personnel on the fireground by means of satellites. The cost for a small department is over fifty thousand dollars. The life of a firefighter is certainly worth more than fifty thousand dollars, but is such an expense necessary? More to the point, can true firefighter tracking be less expensive? The answer to *that* question I believe to be yes.

It's almost embarrassing to say that, at the start of this new millennium, the vast majority of departments have trouble tracking their crews and individual members at fires and other incidents. There are several reasons for this. As always, the T word, tradition, still wreaks havoc within the fire service. At times we hang our hats and our lives on what we consider to be unassailable tradition. The mandatory use of SCBA is still resisted in some departments. All manner of deficiencies are explained away with those most specious words, "That's the way we've always done it."

A second reason firefighters go missing has to do with problem solving. Firefighters are proactive people. If a firefighter sees something that needs to be done, his natural impulse is to do it. On face value, this isn't a bad quality. This tendency leads to freelancing, however, which in turn leads to roaming around the fire scene, correcting little problems as they arise. This is the rolling snowball at the start of an accountability problem.

A third reason traces to our reluctance to speak over the radio. We have been taught that being on the microphone too much causes radio congestion, and this is true. Broadcasting gratuitous, rambling, overly verbose reports is certainly a waste of airtime. Informing the IC that you're moving your search from Division 2 to Division 3 is not. Period.

Even at this single family residential structure fire, Command must continually know the whereabouts of all on-scene crews. *All Photos this chapter Credit Ron Jeffers*

The true goal of accountability should be to track the location of crews on the fire scene, both inside and out. My problem with most of the accountability systems in use today is that they only let the incident commander know who is in the building and who isn't.

If I'm the IC, standing outside a three-story warehouse fire, and the structure partially collapses or the fire acts up and flashes over in a particular area, I want to know two things. First, are there crews in that area? If there are no crews present, then I can react to the problem, but at a different pitch than if members are in trouble. Second, if there are crews in that area, how many am I looking for? When a rapid intervention team informs me that they're bringing two members out of the building, can I relax or am I still missing two more?

Keeping track of the crews on the fireground, both inside and out, can be accomplished without tags, boards, and computer chips. It can be done at little or no cost to the department. What it takes is the realistic and dedicated use of the incident management system, as well as the commitment of company officers.

Many different types of accountability systems are in use today. To my way of thinking, that's part of the problem. What we use in Toledo isn't compatible with the systems of other departments, with whom we might respond in mutual-aid scenarios. If we need a rapid accounting of all of the on-scene crews, we could be in serious trouble. Some

Ten minutes into this three story warehouse fire, would you have the ability to track the location of all interior crews?

departments would be leaving accountability boards at the door to the structure, and others wouldn't even have a system in place for knowing what units responded, let alone who responded on each unit!

The system we use in Toledo employs acrylic boards of approximately eight inches by ten inches. Each officer writes on his board the names of all of the members on the apparatus on his tour. The boards are left in the cab of each apparatus until the incident commander asks that they be brought to the command post at the sounding of a second alarm. The senior battalion chief also makes an accountability or riding list at the beginning of each tour, identifying who is riding where on all companies in the city. This information is then faxed to the on-duty battalions, to my office at headquarters, and to my home fax. The battalions carry this riding list with them throughout the tour, and I keep one with me in the event of a multiple-alarm fire. At any incident, we will have, at minimum, a sheet that identifies who is on duty and on what apparatus each member is riding.

The Fire Department of New York uses a daily roster system to identify on-duty members. Fireground procedures and aides at the command post keep track of the crews. If a collapse or flashover were to occur, the names of the missing members would be identified. Still, some FDNY officials admit that they have a hard time tracking the exact whereabouts of crews inside structures.

Some departments use a passport system. This is designed to provide for a much more controlled accounting of where crews are on the fireground. In order to move from one area to another, each officer must take a passport with him as he moves about the scene. When he arrives at the next destination, he hands his passport to the officer in charge of that specific area or entrance. Such a system requires discipline and the cooperation of the officers at the scene. Still, it lacks one vital aspect of accountability in that it fails to track crews on the fireground or, even more importantly, inside the fire structure.

At this Lodi N.J. occupied commercial fire, crews entering mst be tracked by Command. All crews need to be tracked at the Command Post. *Credit Ron Jeffers*

The following is a synopsis of the precepts of NFPA 1561, *Standard on Emergency Services Incident Management System.* The comments are my own.

1. *The incident commander shall be responsible for accountability.* This is a no-brainer. The incident commander has to be the one responsible for accountability. The actual task of accounting for personnel can be delegated if the IC chooses. However, the incident commander is always the responsible party.
2. *The department shall provide a system for rapid accounting of all personnel on the scene.* This can be accomplished through the use of personnel accountability reports (PARs).
3. *All officers shall maintain an awareness of the position and function of all assigned members.* This awareness shall serve as the basis for accountability. Teamwork is at the heart of our industry. We train as a team, drill as a team, and eat as a team. Why don't we fight fires as a team? If we're going to have true accountability, we must strive to keep crews together.
4. *The incident management system shall provide a system for accountability.* Whatever version of IMS you use, it should provide a means whereby crews can be tracked at incidents. If you use pre-incident assignments, then once crews are inside and move to an area where procedures say they should not be, then the IC must be notified. That's accountability.

Even when a first alarm assignment fights a fire in an occupied commercial structure, Command must track all on-scene crews. *Credit Ron Jeffers*

If your department uses incident-specific assignments, then when crews change location, they should be required to inform the incident commander. That's accountability. All of this should be an inherent part of your IMS. You should be able to walk up to an IC at any fire and say, "Where are your crews?" and he should be able to tell you, not "inside" or "outside," but *where*, specifically, in terms of a floor, sector, or division.

5. *Personnel who arrive on the scene by means other than a fire apparatus shall be identified and accounted for.* This edict is predominantly for rural America. Still, departments often have members who arrive by means other than apparatus. I'm one of those members. In Toledo, I arrive in my own jeep, but no matter. Firefighters who arrive in their own vehicles must collect in a specific area. Call that area what you will, whether a manpower pool, a staffing area, staging, or whatever. Deputy chiefs and others of rank should report either to this same area or directly to the command post.

6. *The system shall identify persons entering and leaving the structure.* This can be accomplished by means of radio. No board, rings, or stick-on tags! When assignments are given, the incident commander can assume that those crews are entering the structure. If you want confirmation when they actually go in, then have the officer or person in charge of each crew report their entry over the radio. If you can see them, this may not even be necessary. To me, it's most important that they inform you when they leave. As the incident commander, it tells me a lot when I hear the officer of an engine company report that they're on the outside. The only other item I need is a riding list or roster of exactly who is on that crew.

7. *The incident management system shall include a standard operating guideline for emergency evacuation.* This is easy, but it's also extremely regional and personal. In Toledo, we use emergency traffic and then plain English to tell crews what we want. From that point on, we wait for personnel accountability reports. Some departments use code words followed by sirens or air-horn blasts. Whatever system you use, be sure that you drill with it often, especially if it involves figurative signals.

Standard 1500, *Standard on Fire Department Occupational Safety and Health Programs* has a section relevant to accountability systems. The following is a synopsis.

1. *The department must develop a written standard pertaining to accountability.* Take whatever you come up with regarding accountability and write it into your SOPs. If you simply have members sign on in the company journal in the morning and that's as far as your accountability system goes, then write that procedure down. If you use a passport system with bells and whistles, then write that procedure down, and make sure that everyone has a copy.
2. *It is the responsibility of all members to participate in the accountability system.* Write it down, and then enforce it! It's that simple. If you have a system on paper that no one follows, then modify it to make it workable, then enforce it. Try it at drills. Use it at vacant houses that you've revisited several times and that no longer have any economic value. Test your evacuation signal. Enforce the command.
3. *The incident commander is responsible for the overall accountability of personnel.* Being IC is tantamount to being the captain of a ship.
4. *The incident commander shall maintain awareness as to the location and function of all on-scene personnel.* Crews must communicate all movement that is not obvious.
5. *Officers assigned as tactical management are responsible for their crews.* Sector officers that are assigned as branch officers or sector officers that have more than one crew working under them are responsible for the location, function, and identification of crews working under them.
6. *Officers are responsible for their specific crew members.* Individual sector officers are to be held responsible for knowing the location and function of all members assigned to them, all the time.
7. *When assigned as a company, members are responsible for remaining under the supervision of the officer.* Individual members are responsible for remaining with their crew at all times.
8. *Members are responsible for following accountability procedures.* Commit a set of protocols to writing. Some violations can be considered to be safety violations. Others can be treated merely as minor infractions.
9. *An accountability system shall be used at all incidents.* Track crews at all incidents, even the small ones.
10. *The department is responsible for making the accountability system work.* Any procedure must be comprehensible and enforceable. Its mechanism should begin at the top and work downward.
11. *The standard operating guidelines shall provide for additional accountability officers as the incident expands.* Big fire, big water! Big incident, big accountability! One officer should be able to track the location of crews at a traffic accident. It may take more than one to track thirty crews at a high-rise fire. If whoever has this assignment is get-

ting bogged down, get him more help! I guarantee you that, at any incident of that magnitude, there'll always be one member standing around with his hands in his pockets. Grab him and give him an accountability assignment. Better yet, grab a good person and replace him with the guy with his hands in his pockets!

12. *The incident commander and members who are assigned supervisory responsibility for a tactical-level unit that involves multiple companies or crews shall assign members to track all of those companies and crews.* Whenever you're assigned, you're responsible for tracking everyone under your control. If you can't handle that, then get help.

Some departments may consider these twelve stipulations impossible to accomplish. Departments that already use a passport-type of system may not even blink at them. It's all relative to what you currently practice. The essential components aren't tags, computer chips, satellites, or any other physical item. The only prerequisites are discipline, control, and an effective incident management system.

When designing an accountability system, begin by taking a serious look at your department. What, if anything, are you currently using? How do you currently operate at incidents? Do you use incident-specific assignments or pre-incident assignments? Is yours a small, rural

Crews at a strip-mall fire are tracked by the Incident Commander. *Credit Ron Jeffers*

department or a major, urban department? Think of your last good, working fire. At the height of battle, while you and your crew were crawling around in hell, how confident were you that the IC, operations officer, and RIT knew where you were inside the building?

Regardless of how your department operates (pre-incident assignment or incident-specific assignment), you need to have a system in place that will track crews inside the fire building. Even if you use pre-incident assignments, you need to ensure that if your crews go to new areas, they will inform the IC.

You also need to look at what other departments in your area do, particularly those with whom you respond on mutual-aid calls. I don't insist that you all mirror the same system. I realize that in today's world, a perfect uniformity of systems may be impossible to achieve. Not every department is alike. Still, if other departments come into my community, I must insist that they use a system by which I can continuously track their crews. Discipline is the key. Know your buildings and your outside help. Finally, test your system jointly at a disaster drill. As much as these drills are disliked and dreaded, they are essential for understanding the flaws in the systems. The true essence of training is error without consequence. It's better to give crews a drubbing in a drill than to lose them at a real fire.

Review the standards to ensure minimum compliance. I know that the NFPA standards aren't laws. They're minimum acceptable standards. They define a standard of care. It's hard to justify in court why a given standard was ignored or not fully met.

As mentioned above, a personnel accountability report (PAR) is an announcement from the officer in charge of a crew that all of the members assigned to him have been located and accounted for. When a PAR is requested, the officer must contact every member assigned to him, either visually or verbally. This includes drivers. Once all of the members have been located, the officer should report to the incident commander that he has PAR. When all of the crews have reported PAR, the IC should give a general PAR for the incident. This announcement is time-stamped for documentation purposes.

There are two types of PARs, routine and emergency. Routine PARs are given by a company officer, anytime he and his crew leave the building, until the fire is under control. In requiring this, we always account for all crews working inside a building every fifteen minutes. Emergency PARS are given at the request of the incident commander or the operations officer.

If I were to design a model accountability system that was both effective and economical, it would look like the following.

Accountability System

I. Purpose

To establish a system by which the location and function of all members can be tracked at every incident.

II. Responsibility

A. The department shall hereby establish and continually follow the accountability system described below. All members shall follow this system at every incident.

B. The system shall not interfere with initial operations. The preservation of life and property is still paramount with the department.

C. The incident commander shall ultimately be responsible for tracking the location and function of all members and crews at an incident. The IC is responsible for ensuring that all members follow the system at incidents. Through the incident management system, the incident commander shall ensure that all members operate within this accountability system and the department's IMS.

D. Command shall maintain an accountability sheet or board that reflects the location and function of all crews operating at an incident.

E. Company and sector officers shall keep the incident commander or operations officer aware of their location and function throughout the incident.

F. Company and sector officers shall be responsible for knowing the location and function of all members at every incident.

There will be a large amount of resources required at this Jersey City fire. Command should designate a single officer to track crews at fires of this magnitude. *Credit Ron Jeffers*

G. Firefighters shall ensure crew integrity and constantly keep their officer or supervisor aware of their location. Firefighters shall not roam or stray from their crew unless directed by a higher authority.

H. In incidents where branch or sector officers are used, or officers are responsible for multiple crews and functions, the branch or sector officers are responsible for the location and function of all crews assigned. Individual officers are responsible at all times for the location and function of assigned individuals.

III. Crew Riding Lists and Integrity

Daily, at shift change, the officer shall establish a riding list. Additional ties, rings, or boards can be utilized to identify crews. From this riding list, a departmental daily roster shall be established. The administration shall assure this riding list is distributed to all appropriate individuals.

IV. Levels of Accountability

A. The department shall use two levels of accountability at all incidents. Level 1 will be used at normal incidents.

1. Level 1 accountability will be used when the incident can be brought under control in fifteen minutes or less (one bottle change for members) and the incident commander can track the location and function of all members at an incident without aid. During Level 1 accountability, the accountability rings, boards, etc. can be left on the apparatus as long as the apparatus is no farther than two hundred feet from any engine at the scene that has handlines in the fire structure. If given an assignment and your apparatus is more than two hundred feet from an engine pumping at the scene, then your rings, boards, etc. shall be brought up to the scene and placed in the cab of any engine pumping at the scene. During Level 1 accountability, individual staff members at the scene shall verbally (face to face or by radio) check in with the incident commander. They are not required to place their individual tags in an apparatus pumping at the scene unless they enter the fire structure.
2. As a crew enters the structure, the officer shall state (by radio) their assignment and intended initial location of operation. When the crew leaves the structure, the officer shall conduct a routine PAR and then report (by radio) their assignment and status.
3. Crews shall inform the incident commander via radio when they change location in the structure. Normally this is required only when changing floors. In large commercial structures, sector officers should report changes in their location when moving from quadrant to quadrant.
4. This process shall take place until the fire is deemed under control and SCBA is no longer required. The incident commander is still responsible for the function

and location of crews throughout the incident. This is especially true if the structure has been weakened by the fire or if fires have previously occurred there.

5. When sent back to quarters, officers shall take their boards with them.

B. Level II accountability shall be used at multiple alarm or mutual-aid incidents.

1. Level II accountability shall be used whenever the incident will take more than fifteen minutes (one bottle change) to bring under control, at any multiple alarm, or whenever the incident commander believes that the incident warrants a stronger accounting of on-scene personnel.
2. When using Level II accountability, all accountability rings, boards, etc. shall be brought to the command post.
3. During Level II, the incident commander shall assign an officer to manage accountability. This officer shall work under the Planning section, if one is established. The safety officer shall not be responsible for accountability at the scene. The RIT officer, if a rapid intervention team is established, shall assist with accountability after his initial duties have been accomplished.
4. Crews given assignments in areas where the entrance is remote from the command post shall give their rings, boards, etc. to the officer assigned to that side of the structure. If no officer is assigned there, then the rings, boards, etc. shall be placed in the apparatus closest to that entrance until an officer is assigned. At that time, he will collect and maintain them.
5. When crews enter and leave the structure, they shall report as being under Level II accountability.
6. When crews change location inside the structure, they shall inform the incident commander or operations officer that they are under Level II accountability.
7. When crews leave the structure for a bottle change or at the direction of the incident commander or operations officer, they shall change their bottles and report to scene staging at the command post and await reassignment. Crews shall report to rehab after their second bottle change.
8. Crews reporting to rehab shall report to the rehab officer. The rehab officer shall keep track of all crews in rehab. After crews have been rehabbed, they shall report to scene staging at the command post and await reassignment.
9. Individual members shall leave their boards or metal tags at the command post if they are assigned a task inside the structure.
10. Due to the numbers of members required and the duration of the incident, Level II accountability shall last until the incident is terminated.

V. Rapid Accounting of Members—Personnel Accountability Reports

A. Personnel accountability reports (PARs) will be the basis for the rapid accounting of all on-scene members.

B. Routine PARs will be given whenever a crew enters and leaves any fire or exposure structure.

C. Emergency PARs shall be given at the call of the incident commander or operations officer.

VI. Emergency Evacuation

A. When the incident commander or any member calls for a partial or total emergency evacuation, the following shall occur.

1. The incident commander or operations officer shall ask for emergency traffic over the radio.
2. The dispatcher shall sound the fire tones for five seconds.
3. The incident commander or operations officer shall announce his notice of partial or total evacuation to all units.
4. On hearing the call from the incident commander or operations officer, the involved crews should acknowledge by giving their unit designation only.
5. After the involved crews leave the structure, they shall give PAR and stand by where they exit, outside of any collapse zone, and await instructions.
6. After all crews have been accounted for, the incident commander or operations officer shall report either a scene PAR (entire area) or partial PAR (only those crews involved) and resume normal radio traffic.

Chapter Five Questions

1. What are the three reasons that the author gives for departments having trouble tracking crews at fire?
2. The true goal of an accountability system should be to ______________________.
3. A ____________________ is an announcement from an officer that he has accounted for the location and safety of all members assigned to him.
4. What are the two types of PAR's?
5. What is the time-frame for placing a fire under control whereby Level 1 Accountability can be used?

Questions for discussion

a) How can requiring the location of all crews at all times affect a department's ability to fight a fire and should this be a consideration?
b) Look at your department's current accountability system and discuss the question the author poses: "Do you feel comfortable inside, knowing how Command is tracking you outside?"
c) How can the driver be accounted for in an emergency PAR?

Chapter Six

Managing the Mayday

Hopefully, none of you will ever have occasion to use the information in this chapter. But practice for it, yes! Practice this at least four times a year! Practice! If you practice anything, practice managing the Mayday!

The size of the incident will have no bearing on when a Mayday occurs. In fact, all things being equal, more firefighters are killed and injured in single-family structures than in large commercial occupancies. Firefighting is an extremely dangerous profession. Many factors make it even more dangerous than it was twenty years ago. Synthetic materials and fewer working fires both make our job more challenging than it was before. Managing fires today is also more difficult. Chief officers can no longer just stand out front and watch their crews go to work. Assignments, accountability, and two-in/two-out rules make managing the everyday fire a mentally demanding task. On top of all that, the unexpected call from interior crews will tax even the most experienced chief, no matter what his department or city.

To create order from chaos, a chief or company officer must understand three things in the event of a Mayday. First, he should know what to expect of his crew when the distress call is heard. Second, he should understand what can and can't be done. Third, he should know what to expect of himself.

The first and foremost thing that an incident commander can expect of his crews is mutiny. This may be the most difficult scene that a chief or incident commander has to face in his career. In time of emergency, the natural tendency for a firefighter is to drop whatever he's doing and run to help his peers. Sometimes this leads to a quick and successful rescue. Most of the time, it leads to confusion, chaos, and possibly more casualties. To control these well-intentioned mutinies, you must anticipate them and control them. In learning how to control them, you'll need practice. You'll need a commitment from management to allow drills at simulated and real incidents. Drills tend to be a small obstacle to overcome. Practicing how to handle mutiny at a real incident is another matter. Pick your battleground. Small fires in vacant structures work well. Make sure that the law enforcement in your area knows what you're doing. When the opportunity presents itself, pull out a good firefighter, perhaps even an entire crew, and have him pretend to go down in a safe and protected area. Then watch as the initial call for Mayday rings out. Forcefully but calmly, direct the crews, and then critique them afterward. Explain where the scenario went well and where certain actions might have actually created more problems. If you anticipate mutinies and attempt to adjust to them, a real Mayday scenario might end with neither tragedy nor additional mishap.

Staging, Safety, and the Incident Commander conferring at a working fire. *Credit Ron Jeffers*

Whatever the outcome, whether successful or unsuccessful, after the Mayday has been handled, the on-scene crews are done for that incident. To expect them to function after witnessing an operation to rescue one of their own is unrealistic. Get other crews on the scene to handle the rest of the initial incident, and send the original crews to a station or other area for a debriefing, then send them back to the station to finish out the rest of the tour. Folks involved with critical incident stress management tell us that it's counterproductive to send these members home. They need to be with their peers in their attempt to return to normalcy.

In reacting to the Mayday, Command should consider time. Brand-new turnout gear will last for about five minutes at 500°F. The dirtier the turnouts, the shorter the time frame. If the area where the crews are endangered is in flashover, you'll have little time to lose. A rapid intervention team needs to get in, and hose streams need to be working the area as soon as possible. If used, you may get up to thirty minutes on a sixty-minute SCBA bottle. Still, whatever the size of the bottle, you may need to transport extra bottles into the area if a sustained search or rescue is required. Also, if the RIT is deploying for an endangered firefighter in a structure, you'll need to monitor the stability of that structure. Partial and total collapses can be anticipated in most types of structures except those of reinforced concrete. The time frame before collapse can range from moments to twenty minutes or more, depending on the type of construction, the degree of involvement, and the size of the structural members.

Command should also consider staffing constraints. The rapid intervention team should be comprised of a minimum of four members. If this team has to split into two teams, four is the lowest number that can safely be split. Together, four members can stabilize and evacuate a seriously injured member.

If the RIT is going to seek out a member in an area where fire could be a concern, then you'll need a hoseline to shadow the team. This line will require a minimum of two members, who will either need a new bottle or replacements every fifteen minutes, given that they're using thirty-minute bottles. If two lines or a backup line is required, then the numbers will grow proportionally.

Let's say that the RIT locates firefighters who were missing. The firefighter is pinned by a steel girder on the second floor of a warehouse. The fire is being held in check with 2 1/2-inch lines. The RIT officer asks via radio for a set of air bags and a hydraulic tool. Two firefighters will be required to shuttle SCBA bottles to the area. Two members will be required to bring in and operate the hydraulic tool. Additionally, two paramedics may be needed to administer pain medication and set up IVs. This will require one escort. Eighteen members are involved in this operation: the four members of the RIT and fourteen in support. An RIT cannot stand alone for a long time. Rapid intervention is a labor-intensive function. The first response of the incident commander after the Mayday call goes out should be to order additional alarms or request mutual aid. Staffing constraints must not hinder the safe removal of a downed firefighter.

When a Mayday sounds, the incident commander should remove all nonessential crews from the scene. There are two reasons for this. The first is to stave off freelancing. It'll be hard enough to work the original fire and the rescue without having other crews run into trouble themselves. The second reason is to gather members who can provide "RIT Support". In order to provide crews for RIT Support, we must first identify what crews are essential and what crews are not.

When the Mayday call is heard, the essential crews include the RIT and any hoseline crew in the last known location of the endangered firefighters. Consider any backup crew in the same location also to be essential. Any hoseline that can be used to hold fire away from that area is essential. Ventilation in progress that will help draw fire and heat from the area is essential. Finally, any search crew that is available and conducting a primary search for viable victims must be continued. All other crews are nonessential and should be withdrawn, accounted for, and reorganized as RIT Support.

Once the incident commander has activated a RIT, pulled out unnecessary crews, and set up support for the RIT, he should initiate a PAR. This allows the incident commander to make sure that all crews are accounted for and that the problem isn't even bigger than he realizes. It also tends to provide focus and a short-term goal for the companies being pulled.

Getting them out, accounting for them, and reassigning them can help fight off mutiny, since these members will have neither idle hands nor idle minds during the process.

Consider using PPV to hold the fire. Normally I'm leery of PPV even at normal fires, so this is a pretty radical suggestion. Positive-pressure ventilation can, however, move a lot of air. By the same token, it can also hold a lot of advancing fire. The intention here is to buy time. If PPV is placed in the area of rescue and directed toward the flames, you may be able to keep the rescue area viable. Start the fan on idle and observe the effects. Watch for increased fire in areas adjacent to the rescue. Inform all attack and backup crews as to what you're doing. Have them report on the effects of the fan as soon as possible.

In considering all of your options, don't overlook the possibility of breaching walls. Evaluate the staffing and time constraints, then consider breaching any walls that might be the only barrier between you and the trapped firefighter. Do you know how long it'll take one of your truck crews to breach a block wall? Do you know how to cut through rebar and how long it'll take? Do you know that a hole in a wall of ordinary construction should be in the shape of a triangle?

The third floor fire escape should be a prime concern for a RIT team officer at this tenement fire. *Credit Ron Jeffers*

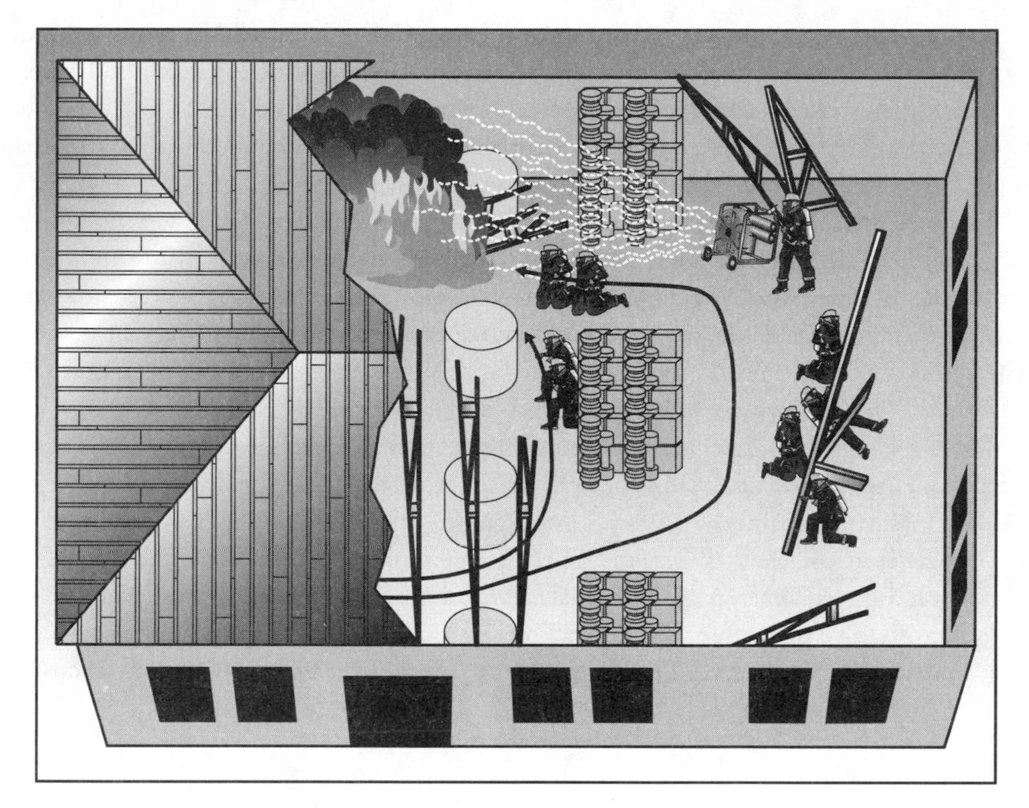

Consider the use of PPV to hold fire while freeing a pinned firefighter

Communicating during a Mayday situation will be challenging at best. Every Mayday scenario that I have witnessed, whether in person or on video, has one common aspect: people are yelling! As an officer, you should put an end to the yelling as quickly as possible. Communications from the incident commander, as well as those from company officers to the crews, should be deliberate, calm, clear, and concise. Only essential communications should be allowed. If possible, designate separate channels for the rescue and the fire.

Remaining calm is part of remaining focused, which means giving symbiotic attention to both operations. While the Mayday response is gearing up, the IC must also remember to deal with the fire. To abandon this effort may spell doom for the missing or trapped firefighters. Since time is so critical, the IC must build a think tank. This is the one situation in which I may appoint two operations chiefs at an incident. One operations chief will stay focused on the original fire. The other operations chief will be in control of the rescue efforts. If staffing is thin, then one operations chief will suffice, but he should have two sector officers under him: predictably, one for the fire and one for the rescue. Whoever they are, the individuals in charge of the fire and the rescue need to be together, and they must talk. In communicating their intentions to the forces outside the command post, three radio channels will help: a command channel, a channel for the fire, and one for the RIT. Even so, the last thing that the IC needs at a time like this is to be burdened with a fireground radio. He should give up his portable and get an aide to monitor the command channel.

The incident commander must be able to handle a multitude of tasks conceptually. He must be able to sort through and prioritize all of the input that he receives. In thinking on two fronts, the fire and the rescue, he may have to deal with simultaneous and conflicting demands for resources. A Mayday situation is certainly one of those times when an IC needs to step back from the moment to weigh requests, then come up with sound decisions based on fact rather than emotion and unrealistic hopes.

There are several considerations in returning to normal operations after the rescue of a firefighter. First, you should conduct another PAR. After that, you should reestablish a plan of attack for the original fire. As soon as you've asked for additional crews to be sent to the scene for relief, reassign crews currently on the scene to necessary sectors. As mentioned earlier, as soon as relief comes, get the original crews on the scene to a debriefing. They should be required to leave the scene. Their adrenaline and focus will be hard to redirect, and subsequent injuries may be incurred if the original crews are allowed to finish out the response.

Chapter Six Questions

1. What are the three things the author believes you must understand in the event of a Mayday?
2. What is the first thing ann IC can expect from his crews in the event of a Mayday?
3. The author suggests the minimum size of a RIT team as _____members.
4. Who are the crews that make up RIT support?
5. What does the author suggest is done with crews at the scene of a Mayday?

Questions for discussion

a) Why would the author suggest that the RIT team stay on the same fire channel as the crew in trouble and that all others go to another channel?
b) Discuss the use of PPV to hold fire while a trapped firefighter is extricted.
c) Discuss the concept of RIT support and the suggested RIT crew/RIT support ratio.

The Fireground

Chapter Seven

Fighting Fires in Older Apartment Buildings

Some of you may start to read this chapter and decide that it doesn't apply to your jurisdiction. Many towns have shot up in areas that were barely on the map when some of the buildings that I'll be discussing were built. By "older," I'll be excluding the more modern garden-type apartments. The buildings that I'll be focusing on in this chapter were generally constructed from about a hundred years ago on up through the 1950s and '60s. Also, many large single-family homes built fifty or a hundred years ago have since been converted into apartments. These subdivided structures pose a tremendous problem from the standpoint of fire protection, and serious life hazards and ease of fire spread are problems that exist in older apartment buildings generally.

An older apartment building can have anywhere from two to dozens of individual apartments. Each unit can be vacant or have three generations of people living in it. Many of these buildings are of balloon-frame construction. Many more have been altered so often that fire spread can be equally rapid, and unpredictable. The biggest threat of fire spread lies in the dominant vertical channel, usually the interior stairway. Fires that find their way to this channel often threaten the lives of all those above and even of those below.

You must get into these buildings to see where fire actually can and will spread. Frequently, EMS runs allow us to get inside and have a look around. While you're waiting for the ambulance, ask yourself how fire might enter the stair shaft. Try to determine how it will affect rescue operations on the upper floors and how best to perform topside ventilation. If you can formulate answers for these questions and incorporate them into your pre-incident planning, you'll have a better chance of success in any firefight that occurs at that address at some later date.

A typical Apartment Building in an industrial city in the mid-west. *Credit Coleman*

Every 5th course in this ordinary wall is a header course which ties the wythes of brick together. Learn to spot these! They signify a lot. *Credit Eric Renzhofer*

Fighting Fires in Older Apartment Buildings

No matter how carefully you conduct your pre-incident planning, the construction features will most likely make or break you at a fire in an older apartment building. It shouldn't come as too much of a surprise to find alterations that haven't been approved. Far too many of these buildings aren't up to building codes.

The exterior load-bearing walls will normally be of either masonry or wood. Buildings with masonry walls will almost certainly have header courses, especially on load-bearing walls. The header courses tie the masonry together, giving it strength. If you find header courses, suspect a true masonry building.

Older wood-frame apartment buildings can either be of balloon or platform construction. The layouts for many of these are similar to those of the masonry buildings. Some of these have the appearance of ordinary construction. Look at the courses. If no header courses are visible, then chances are you'll find wood studs in the exterior walls. Fire can and will spread vertically inside these apparently brick walls.

With ordinary and wood-frame buildings, the interior walls will be of wood. Original walls will be of lath and plaster. Newer walls will be of plasterboard.

The roof assemblies will almost always be supported by wooden rafters. The rafters will be nearly horizontal in flat roofs and at about a 45-degree angle in gable roofs. Most build-

A wood frame apartment in an industrial city. *Credit Coleman*

ings that were originally designed as apartment buildings have flat roofs. The cockloft area will contain the last set of ceiling joists and the rafters above them. The space between may be filled with insulation, but this isn't likely. Some buildings have a scuttle hole to the cockloft and others don't.

The final exterior weatherproof surface, for either type of building, will be of tarpaper; tar and gravel; or of a newer membrane roof over either wooden planks or plywood. If you find plywood, suspect that alterations have taken place on the existing roof.

Buildings of true ordinary construction have their upper floors tied into either joist pockets or a corbel. The floors are supported by 2 X 8 or larger wooden joists. These will generally run parallel to the shortest exterior walls—normally the front and back walls. This would indicate that the side walls, as well as any interior walls that run perpendicular to the joists, are load-bearing walls. On top of the joists will be old wooden floorboards. As with roofs, the presence of plywood in the floor indicates renovation. The set of joists that support the first floor will also be set into pockets; otherwise, they'll be resting on a foundation.

Wood-frame buildings also have wooden floor-joist systems. In balloon construction, the joist on the foundation will be nailed to the exterior side wall. The joists rest directly

An eight room Apartment Building. Do you think this building was originally designed as an eight-room Apartment Building? *Credit Coleman*

on the sill (the bottom portion of the balloon-frame exterior wall, as opposed to the sill in a platform frame, which rests on top of the foundation) and are nailed to the studs. On the upper floors, the joists rest on a ribbon and are again nailed to the studs. The subfloor and floor sit on top of the joists.

Besides the stairway, which is the dominant vertical channel, hidden voids within the walls can also cause concern. If fire enters the interior walls, it can travel to all areas of the building, especially in the case of balloon construction.

When large single-family homes are subdivided into apartments, rhyme and reason are usually left out of the equation. Several small one- and two-room apartments can be crammed into a structure originally designed to accommodate three bedrooms and a bath. Many such renovations only provide one way in and out for the tenants. Often the floor plans are asymmetrical, and dead-end hallways may be present.

Many attics are also converted into apartments. Some of these lack interior stairways. Fire in almost any portion of a building can bring heavy smoke and intense heat to an attic apartment. Consider these areas to be a priority for early search.

Civilian life safety, firefighter safety, and fire spread are primary problems in older apartment buildings. In many large cities, these calls are where the local fire department bread-and-

An ordinary constructed (exterior walls of brick or block) Apartment Building. *Credit Coleman*

butter fires are. For most departments, however, the apartment building fire isn't the most common type of call, so a little extra forethought and planning will go a long way toward ensuring a successful outcome.

Even given a degree of familiarity with building types and construction methods, fire spread can be unpredictable. Opening up early and often will help you detect whether fire is spreading vertically or not. This is where larger departments, using concurrent pre-incident assignments, have the edge on smaller departments, which must rely on sequential or incident-specific assignments. Having several units arrive almost simultaneously and being able to put the second engine company above the fire to check for extension saves valuable time and real estate. Understanding construction features and having sufficient personnel keeps many fires small and manageable.

Cutting off spread and capturing the dominant vertical channel are the most important strategic considerations at these fires. Place the first line so as to keep the fire confined to the original apartment. In almost every case, running the line up the front stairway will prove the most expedient route. Normally, older apartment buildings have center stairways, with the apartments arranged symmetrically on either side. Four units per floor is the norm, though

A Kerany N.J. apartment fire. This Apartment Building is wood frame with a gambrial roof. *Credit Ron Jeffers*

Two floors of apartments are over this storefront in East Orange N.J. *Credit Ron Jeffers*

some have only two apartments per floor. Whatever the configuration, consider the options.

Suppose you're confronted with a working fire in a four-story apartment building. The fire is in the unit on the first floor in Quadrant A. Flames are venting out of two windows on Side B of the structure.

If you initially place the line through the exterior windows, what will happen? If the door to the apartment is open, you'll push fire throughout the apartment and possibly into the hall and up the stairway. What will happen if the hallway door to the apartment is closed? You'll probably kill any civilians still inside. You'll also force heat and smoke into any structural openings within the apartment.

Now imagine that the initial line has been taken through the front door and up the stairs. What will happen if the apartment door is closed? When you and your crew are ready, you can force the door and advance the line against the seat of the fire. The flames, smoke, and heat should be pushed out the window that's already acting as a vent. If the door to the apartment is open, you may have to knock down flames in the hall as you advance toward the seat of the fire.

The dominant vertical channel is the main killer in these fires. The by-products of combustion will travel upward and endanger the occupants on the upper floors. If no vent to the outside is provided, the smoke and heat will bank downward. Two tactics will serve to remedy this problem. The incident commander must assign roof ventilation directly over the stairway, and he must also bring in a backup line. This line should ensure that the stairway is clear of fire and that crews and any self-evacuating tenants can transit the area safely.

If the fire has originated or traveled on the stairs, it may be better to run the initial hose through the rear door of the building, especially if an apartment on the front of the building is involved. The ladder crew or an outside vent man (OVM) should take out the stairway window on the fire floor, and the attack crew can advance on the fire and push it outside.

Backup line with the Attack crew on an apartment fire on Division 2

The attack crew will need a maneuverable hoseline for these fires. A 1 3/4-inch line is normally sufficient for the initial attack line. Fire flows of 200gpm will be required if approximately half of a single apartment is involved.

You'll need to check the apartment directly above the fire for extension as soon as the flames have been knocked down. Smaller departments will either have to assign this task to another responding crew when it arrives, or the attack crew will have to go up for a look.

The ventilation crew's immediate concern on arrival should be with the dominant vertical channel. If staffing on the truck prohibits splitting the crew, then the ventilation tasks will have to be prioritized. There is a definite need to provide an avenue for the attack crew to use in pushing the fire. If a member of the ventilation crew cannot create this avenue, then someone else must render it. Next,in most cases, to save the most lives, topside ventilation should be commenced as soon as possible.

Many of these buildings have a skylight over the main interior stairway, and this certainly makes initial topside ventilation easier. Many such windows are padlocked. A halligan tool can usually make quick work of the lock so that the window may be opened. If the window can't be opened quickly, break it. Notify the interior crews of your intentions before knocking out the glass.

Unless needed elsewhere, the ventilation crew should remain on the roof in the event that fire takes some other avenue and finds the cockloft. Carbide-tipped chain saws, gasoline-powered circular saws, and other power saws work well in opening up the roof and cutting out hot spots.

Horizontal ventilation, besides the initial venting, can be done with windows and doors after the bulk of the fire has been knocked down. Many departments still use negative-pressure electric fans for ventilation. These can be effective in spot-venting small areas.

Positive-pressure ventilation should be used with caution during the early stages of the firefight. Unless you drill with PPV and use it all the time, and unless you're certain how an influx of air will influence fire spread in each individual structure, it may be best to wait until the fire has been knocked down before putting these fans to use. No doubt, PPV will clear heavy smoke conditions rapidly, but unless you're an expert at the procedure, it may be better to rely on more conventional means of removing the smoke.

A four story Apartment Building in Toledo. How many apartments and occupants would you expect to find in an apartment of this type? Pay attention on EMS runs. *Credit Coleman*

Under most circumstances, initial lines will be taken through the front door of these older Apartment Buildings. *Credit Coleman*

Would you guess the interior stairway is directly above the front door by the configuration of the windows on the second floor in this Apartment Building? Know your buildings! *Credit Coleman*

Fighting Fires in Older Apartment Buildings

Many apartment buildings lack smoke detectors. Of those that have detectors, many lack the batteries to power them. This is especially true of downscale residences. The occupants may not have early warning of the fire, and there may be no secondary means of egress. Even if there is another way out, rapid fire spread and heavy smoke conditions in the common hallways may thwart most of the occupants from rescuing themselves.

In commencing search operations, the rule is to start as close to the fire where savable victims might be, then work back and up until the atmosphere is no longer IDLN (Immediately Dangerous to Life and Health). As the search officer approaches the building, he should be reading the building for Search.

The best way in and the area to start the search can often be determined from outside. Remember that the attack crew can usually search the fire apartment during and after knockdown, as well as while checking for extension. This will usually leave between one and three other apartments on the fire floor. If you believe that savable victims might be in these units, then that's where search operations should begin.

Depending on the size and layout of the individual apartments, the oriented search may work very well. A modified oriented search may prove safer and more efficient if the apartments are very large or very deep. You may conduct a modified oriented search by splitting a four-man crew into two teams, with each team searching an apartment. In both instances, one oriented man either remains in the hallway or no deeper into the apartment than the first room. This oriented man remains in voice contact with the searcher. When the search has been completed, the two crews either move down the hall or up to the next floor. If the search crew only has three members, then one member becomes the oriented man while the other two search the same apartment. The key is in maintaining voice contact with the searchers; if this can be done while one member searches an entire apartment, then you may use the regular oriented method.

Backup crews play an important role in fires in older apartment buildings. One of the primary concerns is in maintaining the most direct route inward for firefighters and the most direct route outward for evacuees. It must be stressed that the sole purpose of the backup group is to protect interior crews and ensure the viability of a safe path. Backup crews do not seek out and extinguish fire. If a crew assigned to backup is allowed to look for extension in the unit above the fire apartment, then where would this crew's focus be?

There could be fires at which the first assignment made is backup. In such a case, the backup line should be placed on the stairway. A fire at which a backup group might be designated first is one in which heavy fire is in the lower floors of the building and threatening the main stairway. In this instance, if the first assignment isn't a backup crew, then it should surely be the second.

There could be situations in which the backup group might need to pull and place two lines, as when extension involves two or more floors. In such a case, the supervisor of the backup group should work between the two lines to ensure that the crews are handling the fire and that additional lines aren't required.

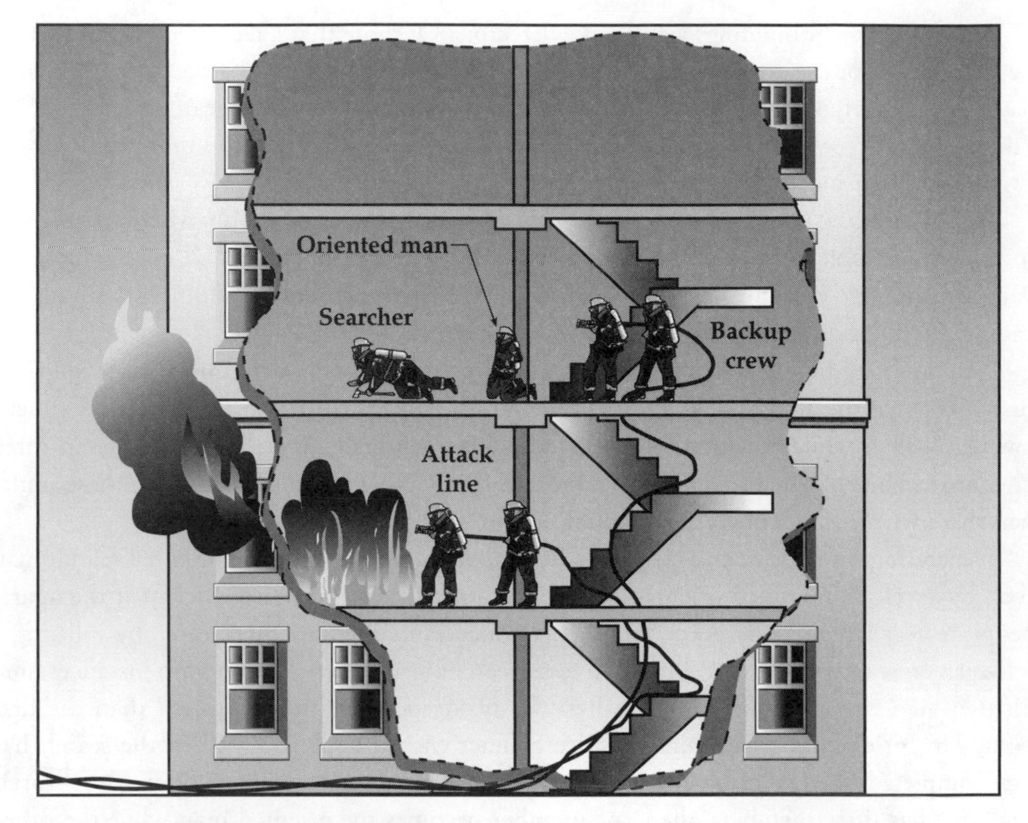

Backup shadowing search on Division 3 while Attack works the fire on Division 2.

In larger departments, checking for extension is usually done according to procedure. Normally, the first place to check is above the fire. Most larger departments proceduralize it so that the second line in (from the second engine on the scene) is taken above the fire. In my department, assignments are prioritized per the incident rather than by running order. Additionally, many smaller departments don't have enough staff to attack the fire, search, ventilate, and maintain a backup group. Such departments usually have the attack group check the floor above for extension after knocking down the flames on the fire floor. No matter how your department assigns its personnel, checking for extension is a must in older apartment buildings.

Most older apartment buildings have walls of lath and plaster. The voids within these walls make prime living spaces for fire. Once fire enters these areas, it can travel to all parts of the structure. Bare hands on walls are still the best indicator as to whether fire has entered the void. Plumbing channels and pipe chases are in line in these structures. Shoddy craftsmanship may also provide an opportunity for extension.

The RIT team and the OVM should pay attention to rear stairways at Apartment Buildings. *Credit Coleman*

When opening up above the fire, don't hesitate to check the spaces around and above the bath and kitchen areas, where pipes are present. Open up the cabinet under the sink to make sure that flames aren't coming up through this space. Check soffit spaces also. If you're unsure as to whether or not a given wall contains fire, check it again with your bare hand in a few minutes. These fires require patience in overhaul. To leave after the visible fire has been extinguished is only to risk having to return an hour or so later to snuff out a rekindle.

In larger departments, the rapid intervention team will be comprised of members trained in the same department as the interior crew. Many smaller departments, however, must rely on mutual-aid personnel to make up the RIT. In many mutual-aid agreements, this is automatic. Regardless of your department's policy, use a RIT in any fire in a commercial occupancy or as requested by the incident commander.

At a working fire in an older apartment building, the members of the RIT need to pay particular attention to the access routes in and out, as well as how the construction features of the building may affect any rescue operation that they may be called on to perform. Besides the front main steps and rear entrance, fire escapes may be present. If fire takes hold of the stairway, fire escapes or aerials may be the only alternate means of entry. Aerials may be a requirement, especially in larger buildings. It should be the responsibility of the RIT to assess the building and determine alternate, emergency means of ingress and egress. Once these factors have been considered, the members of the RIT can hopefully remain at the command post and concern themselves with accountability and the "what ifs" of the operation. One member should constantly review or even maintain the accountability board. Logically, there is no better person to track the location of interior crews.

Scenario

The fire occurs in a building of railroad flats, or what some people call old-law tenements. The building is of ordinary construction. All of its load-bearing exterior walls are comprised of brick. The inside surface of each wall has a coat of plaster to give it a finished look. The building is four stories tall and has a full basement. It is actually two structures joined by a party wall. This party wall is not violated by any opening on any floor or in the basement. However, the cockloft is common to both addresses. Both buildings have a single center stairway. There are two apartments on each floor, one on each side of the stairway. They are long and narrow, running from the front to rear of the building. Each is approximately 1,500

The front of the fire building. Four story ordinary "railroad" Apartment Building. *Credit Coleman*

square feet in size, with three bedrooms and a full bath. Each also has a wooden balcony and stairway in the rear. There are no apartments in the basement, only storage and laundry facilities. All of the utility shutoffs are in the basement.

The fire starts before dawn on a Sunday morning. A jealous boyfriend ignites a flammable liquid in the living room of the unit on Division 2, Quadrant A, after kicking in the door. The flames spread quickly through the apartment, as well as into the hall.

Small-Department Response

Naturally, a small department will have to handle such an incident differently than one of larger size. Staffing constraints will be a key factor in the response. The initial efforts must be aimed at knocking down the fire quickly and then venting the building. Tasks such as salvage, overhaul, and cause investigation must be given a lower priority, since halting the progression of the fire will save the most lives.

The initial officer must assume command, and the officer of the second-arriving unit should work with the two crews. This should give him enough manpower to mount an attack.

The rear of the fire building. Note the stairway configuration. *Credit Coleman*

The case study is in this four story railroad type Apartment Building. The fire originates on Division 2.

The rear of the case study. Smoke conditions are tolerable for the OVM in the rear. The staircase will help with RIT's concerns.

With only one or two initial units, attack must be the primary assignment. The purpose of the line should be to confine the fire to the apartment or floor of origin. To do this, the hallway must be controlled. The size of the line should be commensurate with the size of the fire and the ability of the crew to stretch it. In my opinion, it's best to get a bigger line—say, 2 1/2 inches—to place in the stairway than it is to get a smaller, more mobile line in place. The objective of the first line is to control the hallway, then to move in and confine the fire. If staffing prohibits the use of a 2 1/2-inch line, pull the 1 3/4-incher and do the best you can for as long as you can. Hopefully you can hold the fire until more water arrives.

After crews have begun to work the fire, the next concern will be ventilation. An outside ventilation man (OVM) can be very effective for fire on lower floors. For fires on upper floors, an aerial may be the best bet. For the scenario given, you should spot the aerial in the front of the building, which will provide access to the floor of fire, as well as to the roof. If the fire is best vented through the rear, then rear access for the outside ventilation would be best. In that case, the rear stairway is the fastest avenue for the outside vent man.

After knockdown and ventilations, and given that staffing is low, you should perform search. The attack crews, while moving in on the fire, can look and feel for victims. This isn't the time to open walls if staffing is low. After knockdown, leave one man on the line and commence search. Good ventilation should help your efforts. Maintain a keen sense as to where viable victims are most likely to be, and avoid wasting time. Look in areas where people would be in immediate danger but not obviously dead. You can perform secondary searches later. If conditions warrant interior operations and fire conditions are such that crews must operate on multiple floors or over a large area, consider placing a backup line. Position it to protect interior crews, not to seek out more fire. When staffing is low, put two members on this line.

Lastly, assign a RIT as soon as staffing permits. Consider this to be a primary initial assignment. At times, you may be able to give these members a second assignment.

As additional crews arrive, secondary tasks can be started. These will include backup if the fight is still in progress and crews are operating lines on multiple floors. Secondary searches must be done in fires with heavy involvement and in remote areas where the conditions are imminently dangerous to life and health. Walls must be opened above, around, and possibly below the fire—and all the way through the cockloft and down to the basement, if necessary. Finally, you should commence salvage. Remember that a lot of people who live in these buildings don't have renters insurance. All they own is in their apartment. Concentrate on mementos and the like. Clothing can be salvaged later.

Medium-size Department Response

As with the smaller department, the initial efforts must be focused on quick extinguishment. The officer of the first-in unit assumes command, and then the remainder of the crew mounts the attack. As before, the size of the line will be commensurate with the amount of fire and the crew's ability to stretch the line. The operation may depend on the ability of a two-man crew to take a 2 1/2-inch line to the fire floor. A hose of smaller diameter probably may not be effective.

The next-in engine crew will normally be asked to pull a backup line to maintain the stairs or darken an adjacent apartment. If the latter is the case, then another crew must pull a backup line to maintain the hallway while the suppression crews advance into their respective apartments.

When the truck arrives, an outside vent man must be assigned, while the remainder of the crew ladders the upper portions of the building, preferably the roof. It's normally vital that the roof be opened over the dominant vertical channel.

When the next crew arrives, preferably the squad, it should be assigned to search. The two attack crews advancing on the fire can and should search as they go or immediately after knockdown. If not, then the IC must be advised so that crews can be assigned that task in the fire apartment.

RIT must be assigned in fires involving these occupancies. Depending on conditions, this may be a primary initial assignment or a secondary assignment.

As with the small-department response, after the initial concerns of attack, ventilation, and search have been addressed, second-alarm crews can be used for overhaul, salvage, and secondary searches. Remember that all viable and marginal victims must be brought out. This is a task separate from that of the member who locates the victim. To maintain focus and optimize efforts, use additional crews to establish rescue groups.

Large-Department Response

The great advantage that a large metropolitan department has is in its ability to amass substantial numbers of firefighters at an incident—usually enough to handle all of the various aspects simultaneously.

The officer of the first-in engine should assume command if the battalion chief isn't already on the scene. The first and second engines should get water on the fire. With five- or six-man engine crews, three large lines can be pulled and placed quickly. The first line should go in to control the hallway and advance on the fire apartment. The second line should take any adjacent apartments or maintain the hallway. If necessary, stretch a third line to maintain the hallway.

The first truck crew accomplishes three tasks. The outside vent man opens up ahead of the advancing attack crew. A forcible-entry team forces doors, working with the attack crews.

The remainder of the truck crew searches for life and fire. The second truck crew searches for life and fire above the fire floor. These members also open up the areas over the stairway.

The third engine will probably take a line above the fire and check for extension. The rescue company will work with the third-in engine above the fire or, if life safety is extreme, search for life on the fire floor.

The first-in second-alarm truck will provide rapid intervention teams. Second-alarm companies assist with overhaul and salvage.

Chapter Seven Questions

1. The biggest threat of fire spread in an older apartment building is the ___________.
2. For the most part, an older apartment building will be up to code. True or False
3. One normal give-away as to the presence of true ordinary construction is the presence of _______________.
4. The unusable space between the roof and the top floor of a building with a flat roof is generally referred to as a _____________.
5. The wooden beam that supports a floor is called a ___________
6. What constitutes the dominant vertical channel in an older apartment building?
7. If maneuverability is needed in the initial attack line at a fire in an older apartment building, then the minimum size the line should be is _______.
8. The skylight over the interior stairway should not be opened for any reason in a fire in an older apartment building. True or False
9. Where should search efforts start in a fire in an older apartment building?
10. What route should the backup crew take into the building under normal circumstances?
11. Pipe chases offer an excellent channel for fire spread in an older apartment building. True or False
12. RIT teams will normally not be needed in a fire in an older apartment building. True or False
13. Where will RIT stage after their 360 walk around?
14. It should be the responsibility of ______ to assess the building for alternate ingress and egress.
15. Once RIT returns to their staging area, they should concentrate on the _____________.

Questions for discussion

a) The author discusses the dominant vertical channel in older apartment buildings. Discuss the "physics" that make this the "killer" in these fires.
b) Discuss the role of the OVM in a fire in an older apartment building.
c) Compare and contrast the role and focus of Backup and an Attack crew in a fire in an older apartment building.

Chapter Eight

Fighting Fires in Garden Apartments

Garden apartments are generally defined as three-story multiple-residence structures, but you'll come across plenty of two- and four-story versions as well. Row houses, town houses, and some low-rise, low-income housing projects also fall into this category of building.

For the most part, garden apartments display a few characteristic traits. Normally there are four apartments per floor, with twelve apartments per section. Two or more sections may join or abut one another. The entire complex will consist of similar, if not repetitive, architecture. Midsize and large complexes, defined as four or more buildings, can be served by a single driveway and group parking lots. With the exception of town houses and some row houses, the individual living units are usually confined to one floor. Many of these will have a balcony.

As cities grow and available land disappears, there is an ever-increasing need to put more residential properties in less space. More and more people are opting for apartments and condos as opposed to single-family houses. Baby boomers are apt to spend more time at the beach, golf course, and ski resorts than at home taking care of the yard. For these and other reasons, the construction of garden apartments is on the rise. The result is clusters of garden apartments that are densely populated and often deficient in ready means of access.

From the point of view of a fire officer, there are several problems inherent to garden apartments. Typically, the single main driveway is narrow. Hopefully, the local zoning officials have been cognizant of the width of apparatus when reviewing the plans for new developments. The access to individual units is often poor. Many of these buildings sit well back from the main drive and have parking lots at a distance, meaning that you may not be able to spot the apparatus close to the front of the building. This negates the use of aerial apparatus. There's often restricted access to the rear of individual units. Given heavy fire conditions at the front of the structure, ladder operations at the rear may be hampered.

Typical section of a Garden Apartment. Would the number of light meters have any significance to the Incident Commander? *Credit Coleman*

The individual units may be designated by numbers, letters, or names, but sometimes nothing sets them apart from the other units. Unless building codes are strong on this issue, there can be as many ways of identifying individual units as there are complexes in the community. Lax codes may also lead to a water supply problem. Long hoselays from city hydrants located on the main streets may be required. Typically, these occupancies aren't protected with either automatic sprinkler or standpipe systems.

The life hazard and fire load can both be high. These occupancies aren't noted for wide-open spaces. Combustible furniture and floor coverings will contribute to the rapid spread of fire and early flashover.

Regardless of their exterior facade, the majority of garden apartments are of wood construction. There may be some local differences, and the year of construction may indicate that a different style has been used; still, the essential mode of construction is similar to that of single-family wood-frame houses. Some garden apartments built in the East in the '30s and '40s have exterior load-bearing walls of ordinary construction. The wood joists were tied to the exterior walls by a wooden ledger bolted to the brick or by a corbel, which is a brick shelf constructed in the wall itself. In some low-income housing projects, the exterior walls are of ordinary construction with concrete ceilings and floors.

One section of a Garden Apartment. Twelve apartments on three floors. About 900 sq. ft. per apartment. *Credit Coleman*

Most of the wood-frame garden apartments were constructed after balloon framing became more or less extinct, between 1940 and 1950. Although I can't name any, I'm sure that there are balloon-frame garden apartments out there. Still, the majority have platform construction. Newer versions are being made with truss-frame construction, which poses additional hazards to firefighters. Typically, ordinary-constructed exterior load-bearing walls can have painted concrete masonry units as a finished exterior, or walls of brick veneer. Wood-frame garden apartments can have wood lap, vinyl, or aluminum siding, or perhaps brick veneer.

The roofs can be flat and of concrete or wood joists, but flat roofs are the exception and not the rule. The flat-roof assemblies can be beam and rafter, with sheathing and asphalt roofing in older buildings. Parallel-chord truss or engineered plywood I-beams with tar and gravel over plywood is common in newer complexes that have flat roofs. Some low-income housing projects can have concrete roof assemblies with a membrane covering of rubber or tar and stone.

Most garden apartments have pitched roofs, either gable or mansard. Some of the older types use a true gable roof system with wooden rafters as beams and a single ridge board that supports the slopes. The rafters are normally 2 X 6. The ridge board is normally 2 X 6 or larger. Newer roof assemblies are of truss construction. These are traditionally 2 X 4 assemblies. Some 2 X 6 assemblies can be found, but these are rare.

With life safety being the prime concern at a Garden Apartment fire, how many members will it take to search this section of a garden apartment with the expectation of pulling viable victims out?

The floors are generally of wood-joist construction, with two exceptions. In concrete-construction garden apartments, similar to low-income housing projects, the floor assemblies are of concrete. Also, newer garden apartments can have engineered wooden I-beam or truss-beam construction. Still, the majority of garden apartments have wooden joists supporting the floors. On top of these, you'll find 3/8- or 1/2-inch plywood, with a variety of floor coverings above that. In cheaply constructed units, there may be particle board or chip board rather than plywood as the subflooring. These deteriorate rapidly when assaulted by fire.

As a firefighter, the key is to know the buildings in your district. Ask questions and visit construction sites. For existing structures, visit the building officials in your area. Those state and local entities that approve construction plans should have prints on hand or other references, such as Sanborne maps, which contain site drawings indicating the type of construction.

Garden apartments tend to be symmetrical. Apartments on the same floor are mirror images of each other. The reason is that it's economical to have plumbing shafts that run in line from the basement apartment up through the attic, serving both kitchens and bathrooms. The caliber of the drywall taping will have a great bearing on how easily fire travels through these voids. Anywhere that you find channels for utilities of any kind, you'll find a ready-

Access to this apartment building may be difficult. *Credit Coleman*

made avenue for the transmission of smoke and flame. Wooden void spaces in tics, cocklofts, and truss lofts can essentially become small, confined lumberyards that be prone to backdraft. Even if backdraft conditions don't exist, intense fire can take hold the space between the top floor and the roof. The truss roof assemblies in garden apartment ildings tend to be more stable than those in other types of structures due to the number alls beneath them, both bearing and nonbearing. Even though partition walls aren't bearin alls, they still lend some support to the bottom chord of the truss. Ventilation should be do ith caution and only if no substantial fire has entered the truss area.

Depending on age and local building codes, these roof assemblies can have 1/2-in ply-wood or chip-board (or less) as sheathing, supported by 2 X 4 rafters. That's the ba ws. The good news is that the roof will eventually vent itself.

These occupancies use 1/2-inch drywall on the walls, either 16 or 24 inches on ter. These walls can be breached very easily in an emergency. Most building codes call for f valls or separations in the attic, cockloft, or truss loft. Expect this to consist of 1/2-inch wall nailed and taped to the truss or wall assemblies. Unless you can observe the building le it is under construction, you should assume that these fire separations have been pierce ften

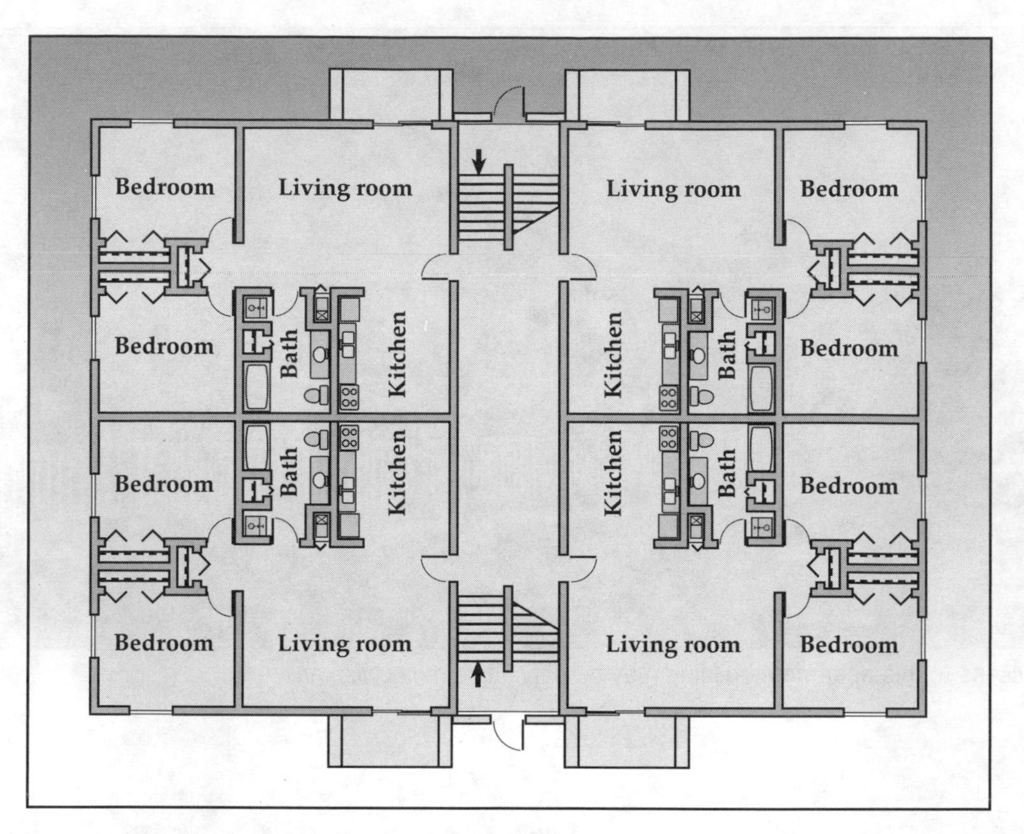

Layout of a typical section of a Garden Apartment.

for phone lines, cable TV, or to add insulation in the attic. It's best also to assume that the area wasn't resealed after the alteration was made.

Like other kinds of multiple dwellings, garden apartments spell a high life hazard. If all of the apartments are occupied at the time of the fire, the potential is for a minimum of twelve fatalities per section. Consider this to be a minimum. Depending on the size of the unit, the season, and the time of day, the risk may be much greater. Some complexes have an apartment or specific area designated as a kind of day-care center. Some sections are specifically designed for the elderly. All of these present additional concerns to firefighters.

Search should be a top priority, and conducting these operations will be labor-intensive in the face of heavy fire and smoke. Using the oriented method of search, it will take a well-trained crew of three approximately five to seven minutes to search two units. This is a maximum and an extremely dangerous tactic due to the size of the apartments. If one crew is used to search an entire section, it will take approximately thirty to thirty-five minutes and at least one bottle change if you're using thirty-minute bottles. This will leave little chance of survival

Rapid ventilation will do several positive things at this fire.

for occupants on the second and third floor to be searched. To be effective in searching a garden apartment complex in heavy smoke, you'll need a minimum of three 3-person crews.

Ventilation may be a concern. If the fire occurs in a single apartment, you'll need to draw the smoke out horizontally and away from the corridors. This is best done with mechanical fans. Positive pressure works best if the seat of the fire has been located and an acceptable vent hole for the smoke is available. If the fire has taken hold of the attic space, natural ventilation by way of a vent hole, either man-made or self-venting, is best. Any attempt to vent the attic must be done with caution. Well-advanced attic fires will, in time, vent themselves.

Sudden collapse isn't a major concern in a garden apartment. Attic areas may collapse down to the ceiling of the topmost floor. These roof collapses may be sudden, but they tend not to bring on total structural failure. Any crews on the roof may be in trouble, but these collapses rarely endanger crews that happen to be working below. Normally, when these structures vent themselves, it's simply by burning off the top. Floor collapse, especially of truss floors or floors on wooden I-beams, also tend to be more localized. Normally, if conditions allow for interior attack, structural collapse won't occur. Certainly, advanced fires that thwart entry can cause general, total collapse. Know what's going on above and below you when fighting an advanced fire that has entered the structural assemblies. Pull the ceiling from below and talk to the crews on all sides of you to find out what's going on behind the walls, ceilings, and floors. The main reason that general collapse isn't a major concern with these types of occupancies, even with truss assemblies, is the presence of bearing and partition walls above and below the trusses. These walls tend to restrict total collapse of the truss, and they add support to adjacent members.

Egress for occupants may be a problem. In Toledo, we were involved with a rash of arson fires involving garden apartments in which the arsonist would set a trailer of gasoline along the interior hallways and light it as he left the building. The occupants were trapped in their units with raging fire in the hall. Even with a normal apartment fire, the stairs and hallways can become heavily charged with smoke and hamper the progress of evacuees.

Balconies can also mean trouble. Normally, occupants who can't flee by way of the interior hall and stairway may go to their balcony. On arrival, you may find trapped occupants screaming and waving from perches all over the exterior of the building. This may be distracting for deploying crewmembers, since many of these individuals on the balconies will sound as if they're in great peril while actually being in little imminent danger. As a general rule, those who are screaming the loudest are usually in the least imminent danger. If the people on the balconies are out of the direct path of flame and smoke, you can put them on your to-do list for incoming crews. Be aware that most balconies are supported by cantilevered beams tied to floor joists. They aren't designed to be a resting spot for ladders or half a dozen

Balconies may be a factor at a fire at this Garden Apartment. *Credit Coleman*

firefighters. For reasons of safety, teams of two and a ground ladder should be the rule for operations involving a balcony.

As always, life safety is your primary concern at a fire in a garden apartment building. Given that, should our initial efforts be concentrated on ventilation, drawing the products of combustion away from those who can't leave on their own? How do you justify not addressing the trapped and endangered occupants on the balconies immediately on arrival? The answer to this dilemma is a matter of staffing. If you don't have enough manpower to conduct search, rescue, ventilation, suppression, and other key operations simultaneously, put out the fire and vent. Look for victims next.

I've had two major fires involving garden apartments recently. In one, forty-mile-an-hour winds blowing through cheap construction literally allowed a fire to pass in front of our eyes. The initial crews made a decision to remove sleeping occupants. Doors were forced ahead of the fire, and by the time crews located and awoke the occupants, they had to crawl under fire at the doorways with victims at their sides. Eight civilians were brought out in this manner. Another half-dozen were taken from windows via ladders. This was a fast-moving fire. We eventually ordered a 2 1/2-inch line put through a window three apartments ahead of the fire. The ceiling was pulled, and the suppression team was told to take a stand there with the hoseline. This tactic worked, and the fire never made it past the 2 1/2-inch stream.

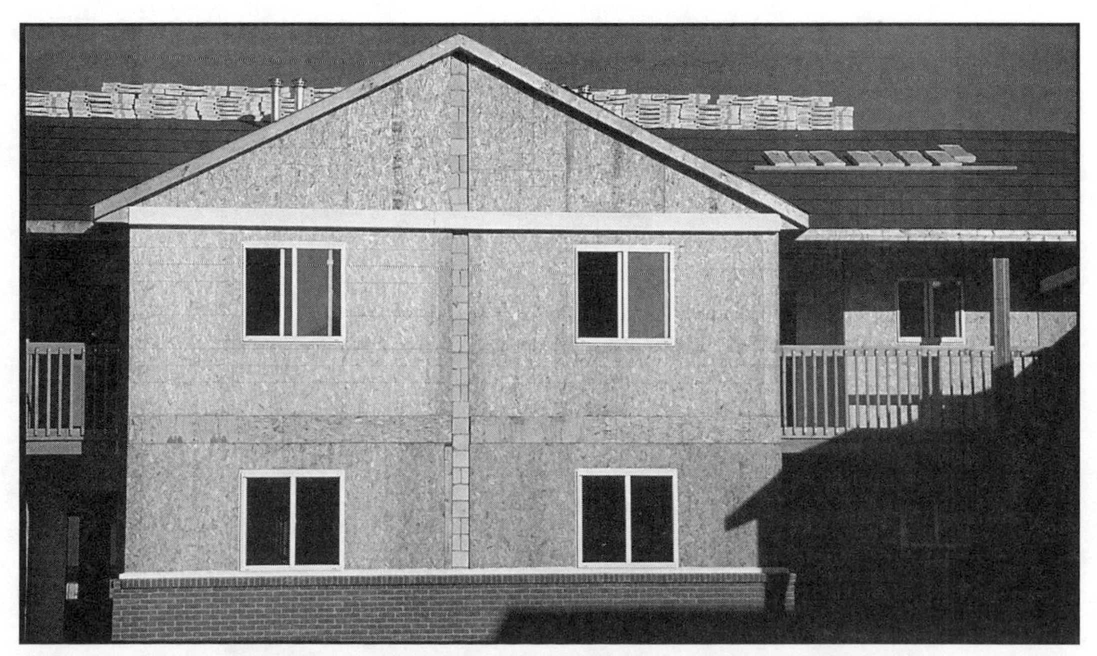

A Garden Apartment under construction. Note the CMU wall that is intended to act as a fire separation between the sections. *Credit Coleman*

In another incident about a month later, we made several rescues, then got ahead of the fire. Two apartments down, we pulled the ceiling, again to stop the progression of the fire in the attic. In this instance, the construction was better, and the fire was stopped well ahead of the precautionary holes in the hall.

The incident commander must ensure that provisions are in place to control any occupants who come out of the building. If left to wander or leave the scene, you'll be hard-pressed to account for their whereabouts when loved ones come to look for them. Avoid headaches. Corral and control the displaced occupants. The information officer is the likely one to handle this strategic concern. He can provide the occupants with information on what to expect, how to contact insurance carriers, and steps being taken to salvage belongings.

In almost every situation, the primary tactic to use is an aggressive interior attack. Consider lines of 1 3/4-inches to be a minimum even at a room-and-contents fire. You may need a lot of water. If you pull a smaller line, you may not have enough to douse a fire that turns out to be bigger than you originally thought. Take in the lines along the most direct route. Normally this means through the front door to the involved section. Avoid throwing in water from the outside, especially with a fog nozzle. The exception to this rule is when the hallway or both lower apartments are involved. One of the principles of fire attack is to place initial lines with the intention of confining the fire to the smallest area possible. In apartment

At this Garden Apartment fire, where would you expect the first attack line to go?

fires where fleeing occupants may be in the hallways, it may be best to take in the initial line from the opposite direction of the fire. This will do two things. First, it will ensure that you immediately get a line between any possible victims and the fire. Second, all other crews entering will also have a line between them and the fire.

Go in and push out the fire. Make sure that an avenue for the fire to follow has been provided. The outside vent man can easily handle this task. Once the door to the apartment has been located and the line has water, go in and knock down the fire. Hesitation can allow for additional spread, and it may cost lives or cause injury. Once the fire has been knocked down, attack crews should move to other known areas of involvement or begin to check for extension. Aggressive attacks of sufficient size are the key. In most instances with fires in garden apartments, you should go in as soon as possible, and go in big!

When the fire is advanced, search can be difficult in garden apartments. Give priority to the involved section and then adjoining sections in the direction of travel. If conditions indicate that there are savable victims in the fire apartment, start the search there. If additional crews become

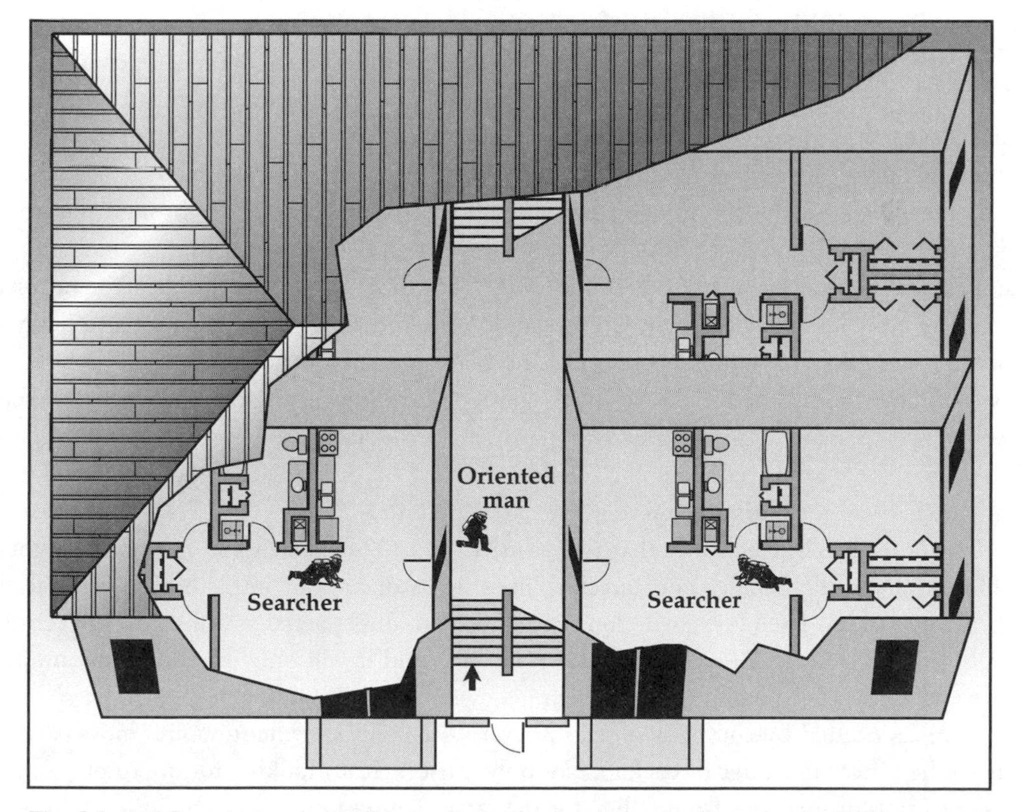

The Oriented Search crew (one officer and two searchers) search a floor of a Garden Apartment.

available, search the remainder of the units on the same floor. If staffing is tight, begin with the apartment on the same side of the hall, then cross over to the other two. Next, search the floors above the fire, beginning with the closest and moving farther away. After all of the floors above the fire have been searched, you can go back down to cover any floors below the fire. I believe that the best method to use is the oriented method of search, in which one oriented man remains in the hall and keeps track of his crews. These apartments are usually small enough that a single member can search each one, using the oriented method. If the IC sends in additional search teams, the oriented man can direct the officer to unsearched areas in the section. In drills conducted in an abandoned hotel in Toledo, we determined that the maximum number of members that one oriented man can effectively control is four. Any more than that and the oriented man has to move around too much and can't track his searchers.

It may be difficult to know at first how many victims might be in the apartments. Some of the units may be vacant. Some may only have one occupant. Some may have five or more. The manager of the complex may be of some help, if he or she is on the scene. As soon as practical, get someone to the rear of the structure. Often a quick accounting of fire concerns in the rear can help you calculate where people are apt to be throughout the building.

Ventilation for fires on lower floors can normally be handled best by horizontal techniques using positive-pressure fans. In Toledo, we like to make sure that we know where the seat of the fire is and that it has been darkened down prior to turning on the fan. If the attack crew hits the fire in a few seconds, then we can begin PPV in a relatively short period of time. Of concern with PPV is that, during a coordinated search, the doors to adjacent apartments must be kept open while the search is in progress. This may negate some of the effectiveness of PPV. However, if all of the other doors are kept closed until necessary, these effects can be minimal.

If the fire is on the top floor or has entered the structure itself, as through plumbing voids, then vertical ventilation may be in order, if your assessment leads you to conclude that the gain will be worth the risk. Opening up the roof directly over the fire apartment, or as directly as possible, will work best. Crews should vent and then get off the roof as quickly as possible. The roof of a garden apartment is no place to take a break after cutting a hole. As crews are opening up the roof, have attack crews pull ceilings in the fire apartment, directly over the fire apartment, and in adjacent apartments to look for extension in the attic space. If the fire is more advanced, you may consider opening up over adjacent sections also. Do so near the fire wall to minimize spread if fire enters that area.

If your assessment indicates that roof ventilation isn't safe, then you may either fight it from below, pulling ceilings and directing lines upward, or wait until the fire vents itself. When the roof has been breached, don't be tempted to direct aerial streams into the ventilation hole. This will do little to knock down the fire, and it will only jeopardize any interior crews and cause additional water damage.

Focus is needed by company officers at every fire. This statement applies most particularly to the officer in charge of backup. Not only must he resist looking for fire to put out, he must also restrain his crew from falling for the same temptation.

Where would you expect the search crew to start and end their search at this fire?

A flat roof on a two story Garden Apartment. Again note the CMU fire separation that divides sections of this complex. What are the chances that this CMU wall has been breached? *Credit Coleman*

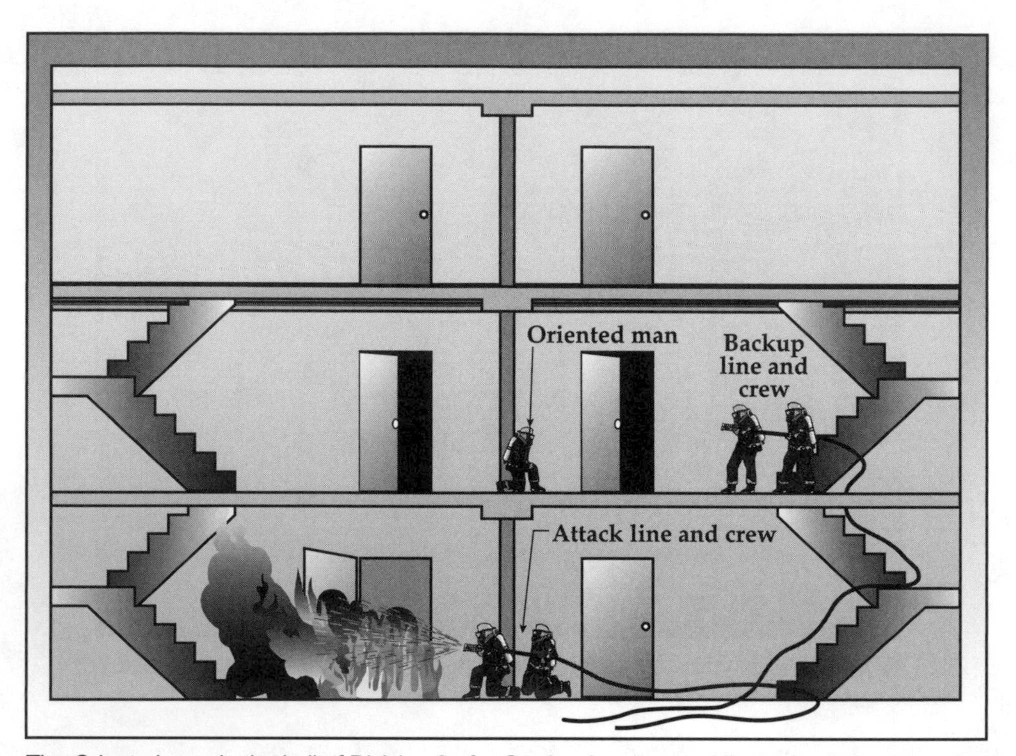

The Oriented man in the hall of Division 2 of a Garden Apartment while the backup crew maintains a line in at the stairs.

When a fire occurs in a garden apartment and you've been assigned Backup, your sole focus should be on the safety of interior crews and providing a constant, unimpeded route in and out of the building. This route should be established for civilians as well as for firefighters. If the attack crew can focus on putting out the flames, the search crew on locating trapped victims, the rescue crew on removing those victims, the ventilation crew on removing the products of combustion, and the backup crew on maintaining the safety of persons inside the building, then the whole extinguishment picture of the fire should come together.

Backup at a garden apartment fire is a relatively simple assignment. Under most conditions, the backup line should be positioned in the same stairway as the attack line. Rarely should the backup line pass through the door of the involved apartment, however. The officer assigned to backup can follow the attack line and check on the safety of the interior crew. Under normal circumstances, the hallway will be the most important area where fire spread must be checked.

When search is assigned, the backup line should shadow the search crew. If the fire is on a lower floor and the search team is on an upper floor, then the backup line must be stretched

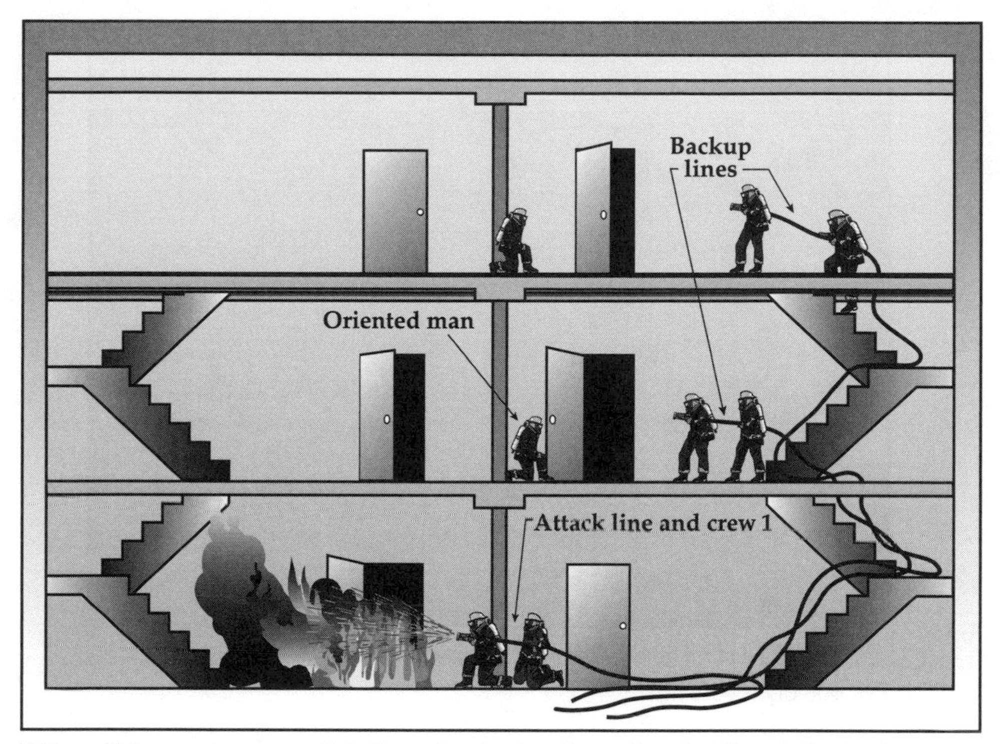

With multiple searches on multiple floors, two backup lines will protect Search the best.

up to where those members are working. Under most circumstances, it should be positioned at one end of the hall, at the stairs, with sufficient slack in case advancement is required. There are few reasons to stretch the line down the hall. If the search crew needs to retreat, then the line can be used to hold the fire and protect the searchers as they back down the stairs.

If the back up team needs to shadow two search crews, one on one floor and one on the floor above, then the line must be stretched to the higher floor and with additional slack. The officer and a firefighter must then go between floors to check on the safety of the searchers. This is another advantage to using the oriented method, keeping the oriented officer in the hallway while his crew searches. Communications are enhanced and made easier with the search officer in the hall and the backup officer on the stairs.

If searches are being conducted on all three levels of a garden apartment, then it's almost a certainty that two backup lines will be required. It's difficult for one officer and a single line to protect all three floors of a garden apartment, especially if all three floors are involved in fire.

After all of the searches have been completed and the fire has been deemed under control, the crew assigned as backup can be reassigned to overhaul or other such tasks.

The front view of a fire in a Garden Apartment. What will RIT be looking for on this side of the building?

For the most part, garden apartment buildings are all constructed the same. They're made with a lot of wood and drywall. The walls will hold in a lot of heat, since drywall is very endothermic, but once fire reaches the internal structure of the building, it can travel widely. When checking for extension, you'll find the walls to be hot. These aren't lath-and-plaster walls. Once drywall is subjected to excessive heat, it begins to give up its moisture. When this happens, the paint on the walls will blister and never look the same again. Don't hesitate to open up the walls. Most of the drywall will have to be removed, anyway. Open them up high, since that's where the heat will be. Pull the ceilings with hooks and pike poles. If the fire has extended to an area of plumbing channels or soffits, open up the area for a good look. Check it several times before you return to the station.

The function and preparation of a RIT in the context of a working fire in a garden apartment will be similar to that of other scenarios. Access in a garden apartment is in some respects easier than it is in older apartment buildings. Most garden apartments have identical stairways and doors, whether front and rear or side to side. Balconies may or may not be present, but most units have windows that can be accessed with ground ladders. Some sublevel apartments have small windows that create a four- to six-foot drop inside. Attic ladders may be required to reach the bottom safely.

The rear view of the same fire. RIT's concerns are different back here.

The extent of fire upon arrival and the fire's location will probably play the greatest role in what should be done first at this fire.

A 360-degree walkaround should help the RIT determine the best location to stage. If the fire is in the front of the apartment, it may be best for the RIT to stage at the rear, where access will be easier, safer, and probably faster. In these situations, the RIT may want and need a separate backup line and additional personnel to stage with them.

Scenario

The fire occurs in a twelve-building garden apartment complex. A single private drive provides access to the twelve individual buildings in the complex. Water to the complex is provided by a city water system consisting of an eight-inch pipe feeding from a twelve-inch main in the street. Hydrants are spaced at approximately five-hundred-foot intervals throughout the complex.

The buildings are about seven years old. Each building has three floors of wood-frame platform construction. Every above-grade unit has a balcony. The walls consist of drywall on 2 X 4s, the floors have 1/2-inch plywood subflooring on 2 X 8 joists, and the roof sits atop 2 X 4 trusses. Each building is comprised of two sections, with fire separation in the attic

The fire apartment from side "A" view. *Credit Coleman*

between the two sections. At the time of the fire, it's unknown whether the separation has been compromised.

The fire starts in a below-grade first-floor apartment when a pet knocks over a halogen lamp. The lamp falls onto a pile of newspapers next to the living-room sofa. The fire spreads quickly. The smoke detector wakes the occupant of the apartment, who attempts to flee and is overcome just as he opens his front door.

Small-Department Response

The first officer on the scene establishes command and assigns the remainder of his crew, along with the second-in unit, to attack operations on the fire floor. This is done on the premise that, when staffing limitations prohibit simultaneous attack, ventilation, and searches, the primary order of business is to put out the fire. A small room-and-contents fire can be handled with a 1 3/4-inch line, but anything involving a space of five hundred square feet or more becomes a toss-up between the 1 3/4-inch and the 2 1/2-inch line. Maneuverability is required, but only after knockdown, and you can't knock down much more than a 20- X 25-

The front view of the case study. Heavy smoke and fire conditions are present on the lower level of this three story building.

square-foot space with a 1 3/4-inch line set at 175gpm or greater. When in doubt, pull the bigger line. Take it in dry, then charge it. After you douse the fire, wye off a smaller cleanup line. This tactic is essential for smaller departments. The purpose of the first line must be to confine the fire to the smallest area possible. Keep the fire out of the hall and then push it out of the apartment.

Once sufficient crews are present, ventilation is the next concern. Take out windows with an axe or a pike pole in the fire room. If the fire is on the upper floor and crews are available, a ground ladder set forcibly in the window will serve a dual purpose, creating a ventilation portal and providing a secondary means of egress.

If smoke is affecting other apartments in the section, then the ventilation must be more aggressive. Positive-pressure or topside ventilation is in order. For fires on lower floors, PPV works the fastest. If you open the roof, vent over the main hallway. For a fire on the top floor, venting over the fire area and pulling the ceiling under the vent hole is the best configuration. The members of the first-in ladder crew can normally handle this task on their own unless excessive fire has entered the attic. Normally, in that situation, the roof will vent itself.

In smaller departments, search will be the last critical assignment to make. If staffing allows attack, ventilation, and search all to take place concurrently, then the search crew

The rear view of the case study. Heavy smoke and fire conditions are present on the lower level of this three story building.

should take the same avenue into the building as the attack crew. While the attack crew takes on the fire apartment, the search team should enter the other three apartments on the same floor. If the search officer believes that none of the victims there would be alive, then the searchers should begin on the floor above. In such a case, a backup line is almost always a necessity.

Rescue groups play a key role in fires in garden apartments. They allow searchers to cover their assigned areas without duplication of effort or wasting time. After the search team has been assigned, a rescue group consisting of a minimum of two firefighters should be selected and located just outside one of the main entrances. The best doorway to be in is one with few hoselines and ready access to the ambulances. When search locates victims, the rescue group of two firefighters can move up to and remove the victim while the searcher continues to cover his area.

Once the primary tasks have been accomplished, the secondary tasks of salvage and overhaul can be commenced using second-alarm crews, if available. The potential for escalation in one of these structures is so great that a quick call for help is almost always warranted. I can't think of too many good reasons not to strike a second alarm at a working fire in a garden apartment building.

Medium-size Department Response

The first-in officer must assume command and initiate attack, ventilation, and search. Most times, if the staff consists of four members, the first-in unit can pull an initial attack line into the building, using two members. This leaves one member to operate the pumps and the officer as Command. Some departments always have the first-in engine secure a hydrant. Others allow the first engine to go up to the building and have the second engine supply the first. No matter how it's done, the water supply must be established, even with a small fire.

I would expect the first hose inside to be a 1 3/4-inch line as a minimum. The placement of this line must be to cut off the fire from the hallway and then to push it horizontally out of the structure. The next should be a 2 1/2-inch line as a backup. If the attack line can't darken down the fire within sixty seconds, an additional attack line will be necessary. This can probably be pulled by the second-in or third-in crew.

Depending on the location of the truck crew at the time that water hits the fire, the truck crew or a member already on the scene must perform outside ventilation, using balcony doors or side windows to render an adequate vent hole.

After outside ventilation has been completed, venting using PPV or natural means must be accomplished. As mentioned above, PPV provides the quickest means. Attack or overhaul crews must check and open up any possible avenue for fire travel near the fire if PPV is used. If negative-pressure fans, nozzles, or topside ventilation is used, the need to open up becomes less critical, but it's understood that these methods are less aggressive and slower. It all depends on your concerns. If you think that there are still victims hidden by heavy smoke, use PPV. If the fire is small and the smoke is light, I'd opt for less aggressive ventilation.

Backup lines are a luxury that medium- and large-size departments can afford to have. Consider using them whenever the involvement on one floor is heavy. A backup line is a must when the fire involves the hallway or more than one floor, and also when search crews must go above the fire. Unless they have to move in to pull out a crew, the backup crew should maintain a position in the hallway, mirroring the search or attack crews.

Large-Department Response

The first-in engine should always secure a water supply and pull a line to the fire. The second-in engine should assist with the stretch and then pull a second attack line or, if conditions are severe, pull a backup line. Most large departments have the manpower to pull a big line, if necessary.

As the initial lines are being deployed, the third-in engine will, in most cases, get a line above the fire. The size of this line will be commensurate with the potential for involvement of the upper floors and truss loft.

The first-in truck crew should provide outside ventilation, forcible entry, and initial search. The second truck crew should ladder the building, perform topside ventilation if necessary, and assist in the search. Squad crews can help pull lines above the fire, assist with the searches, and establish a rescue group. If second-alarm companies are required, they can perform salvage and overhaul. They can also relieve the initial crews inside.

Chapter Eight Questions

1. How many apartments are normally found on one floor in a single section of a garden apartment?
2. What usually prohibits the use of an aerial at a fire in a garden apartment?
3. Generally, garden apartments are ____________ construction.
4. The majority of roofs in garden apartments are supported by __________.
5. Floor plans in garden apartments are _______ images of each other.
6. What is the rule that the author used for the number of firefighters and ladders permitted on a balcony in a garden apartment?
7. Consider a _____ line as the minimum initial line to be used at a fire in a garden apartment.
8. What method of search works well in a fire in a garden apartment?
9. If there is extensive fire in the apartment involved, where should the search normally begin in a garden apartment?
10. If fire and smoke enter the halls of a garden apartment, where should the roof be vented (if you choose to vent the roof)?
11. Under most circumstances, the back up line should be positioned in the __________.
12. If a search group is assigned, then back up should shadow __________.
13. Drywall gives up _____________ when it is heated.
14. A _____________ will assist in providing access to the lower level of a garden apartment.
15. What should the minimum size of the RIT team be at a fire in a garden apartment?

Questions for discussion

a) Discuss the pluming voids as routes of fire and smoke travel in a garden apartment.
b) Discuss the use of a Rescue group in an advanced fire in a garden apartment as opposed to having "search" remove victims.
c) Discuss the pro's and con's of PPV at a garden apartment fire.

Chapter Nine

Fighting Fires in Strip Malls

Strip malls are a phenomenon of the 1970s, '80s, and '90s, much as taxpayers were in the East around the early 1900s. To put it another way, the strip mall is the modern taxpayer, since they're both predicated on essentially the same concept. Old taxpayers were generally disposable buildings placed on property purchased for later speculation when land values rose. These cheaply constructed buildings were generally two stories tall, with several occupancies divided by party walls. The rent from the store owners would help pay the taxes on the property, hence the name.

Although no one can attest to the motives of every strip-mall owner in the country, one can note similarities between these buildings. Generally, they can be defined as one- and two-story lightweight buildings that house several different businesses under a single roof. Although rare, they can have residential occupancies on the second floor, but I know of none that have residences on the first floor. A few have basements. Most are constructed on concrete slab floors.

Early collapse is the key concern at a fire in a strip mall. The manner of construction and the fireloads within are both at fault, and the type of occupancies present can vary widely. A strip mall can house anything from a beauty salon and a restaurant to a carpet store and a laundromat. Some police substations are even located in strip malls (especially those located near donut shops!).

Access to the rear is always a matter of uncertainty. Expect steel doors, security bars, and concrete-block walls. Access through the rear may be difficult at best. Glass is comparatively cheap. If you need to get in, take out the glass in the front.

Normally, life safety isn't a key concern at fires involving strip malls. Fires that occur during business hours are noticed early, and those that occur after hours may not pose any life hazard at all. It's rare for there to be a night watchman on the premises. Still, these

A new Strip Mall with a flat roof. This Strip Mall contains a restaurant and two beauty salons among other occupancies.

An old "tax-payer" in an old part of the city. These have evolved into today's Strip Malls.

The fire in this Strip Mall is extensive. With this amount of smoke showing, aggressive initial interior attacks are warranted.

occupancies can hold surprises. What may seem like a mundane office building with a confusing name on the facade may actually be an abortion clinic or a research lab for who knows what. Expect the unexpected. Explosives, hazardous materials, and caged animals may all be found in these occupancies.

There isn't normally any built-in automatic fire protection in a strip mall. Standpipes are virtually nonexistent and, unless they're required by code, automatic sprinklers are rare. Some have central-station fire detection. This will help in early detection, but you should expect to find an advanced fire in a strip mall if you happen to be responding after business hours.

Don't let developers fool you. These buildings aren't designed to take much more than strong gusts and an average snow load. Fire probably wasn't a word in the architect's vocabulary. There are probably no engineered extras built into the structure to hinder smoke and flames. A strip mall is only there to make money until something more profitable comes along.

The exterior load-bearing walls are generally of ordinary construction, with concrete masonry block being the customary material. Some strip malls have brick veneer walls tied to the block. These combination walls will generally be on the visible sides of the building, and those sides that are normally out of view will be of concrete masonry units only.

A few strip malls have exterior load-bearing walls constructed of tilt-up concrete. These normally sit on concrete slabs, and the roof assemblies will be of concrete also. These buildings are tied together with steel rebar, and they're only as strong as the steel, which will begin to fail at a mere 800°F.

A few strip malls have wood exterior load-bearing walls. Those that do are usually on concrete slabs. These wooden buildings tend to be comprised of the lightest materials possible that can meet building codes and manage to hold up the structure. Anticipate early collapse.

The quality of the roof assemblies varies from very sturdy to quite flimsy. Some are so flimsy that you'll feel a spongy sensation when walking on them, even without the effects of fire. Wood-truss and steel-bar joist are probably the two most common types, in that order. As a group, wood trusses lack mass, and there are several varieties in use. Flat roofs use parallel wood trusses or plywood I-beams as rafters. Plywood or particle-board sheathing is then placed on the rafters, and a weatherproof covering is applied over that. Pitched roofs are generally made of triangular Fink wood trusses. Occasionally you'll find ordinary wooden rafters of anywhere from 2 X 6 to 2 X 12 inches. Still, strip malls with real wooden rafters are generally older and quite rare. Pitched roofs on strip malls are usually finished with asphalt shingles.

Access in the rear of a Strip Mall can prove time consuming in the "heat of the battle".

Steel-bar joists are used to support flat roofs. These will generally be covered with a decking material plus a weatherproof coating of tar and gravel.

If you're unsure as to the composition of the roof of any strip mall, treat it as though it were being held up by trusses. If the cockloft (flat roof) or the truss loft (pitched roof) is involved in fire, let the roof vent itself. Given the weight of the construction materials, you may be able to open up holes using pike poles from an aerial platform, but this will be a slow and tiring process. The roof will soon open up on its own.

At most incidents where fire has entered the truss area, it isn't a good idea to place crews on the roof to ventilate. Pitched, flat, and shed-style truss roofs can all collapse suddenly, plunging the vent crew into the fire below. Steel gusset plates are a poor material when it comes to holding two truss members together in the face of fire. The connecting points only embed 3/8-inch into the wood, and the steel bends when heated. The failure of one part of a truss weakens the entire assembly, and trusses can fail in series, like dominoes.

Of all the assemblies described above, only the true rafter-supported assemblies can be trusted for topside ventilation. These roofs rarely experience general and precipitous collapse. Rather, they tend to sag and become spongy when weakened by fire. The last comments are not indicative of truss roofs.

As mentioned above, the floors of most one-story strip malls are of poured concrete. Generally, "poke-throughs" and utility raceways present the only concern for extension. Those strip malls that do have basements usually have wooden floor assemblies. Older buildings will have two-inch floor joists, with 2 X 6s and 2 X 8s being the norm. Depending on the span, there may or may not be columns or girders for additional support. Some floors may be supported by load-bearing block walls in the basement.

Newer two-story strip malls normally have wooden truss or I-beam joists that support the second floor. Older malls have true two-inch solid joists. If the mall is built on a concrete slab, then the ceiling area and the truss loft are the only place where fire spread will be abundant.

The strategy for fire involving a strip mall is an aggressive large-caliber frontal attack in conjunction with aggressive exposure protection. This may sound simple, but speed and coordination are key. A strip mall is essentially a closed container with excessive fire loads and a lumberyard for a lid. In order to stop the progression of fire, speed is essential, and deck guns and 2 1/2-inch handlines are the tools of choice.

To achieve the quick knockdown, it's best to attack from the front. Some cringe when I say to take the window, but if we leave no building to hold the window frame, what good is saving the glass? Compared to walls and a roof, glass is cheap. Don't get sidetracked into forcing the back door if the fire is in the front. For those who advocate attacking from the rear if the fire is in the front, expect a delay that, in most cases, won't be worth the savings. Normally, there will be no savings.

As soon as lines are in operation, force entry into the nearest exposures. Once again, take out the glass. Don't wait for a key if the fire in the original unit is advanced. Seconds count. Take the glass, then pull ceilings at the party wall. If fire is already in the first exposure, then

A two story Strip Mall. Would you expect the upper units to be second floors to the stores below or individual units. Know your buildings!

This Strip Mall has a pre-cast concrete roof assembly. Better take several saw blades if you intend to vent this roof. We had a second alarm in this Strip Mall a year ago and the roof never got opened.

The National Fire Academy's fire flow formula will indicate approximately 900 gpm initial flow. (44' x 60' = 2640 ÷ 3 = 880gpm)

go to the next and take the same actions. Your objective is to get ahcad of the fire up top and confine it to the smallest area possible. Don't play catch-up. Get to where there is no fire, then work backward toward the original occupancy.

Strategically, that's it for a working fire in a strip mall. Use heavy-caliber streams via the fastest avenue possible, then pull ceilings in the exposures and stop the progression of the flames.

Forget the roof for the most part. These fires can be very well advanced on arrival and, due to great fire loads, large quantities of water are required for quick knockdowns. Remember the NFPA's fire-flow formula: length X width ÷ 3. Suppose that the average occupancy in a strip mall is 60 X 60. That's a good-sized shop. According to the formula, 60 X 60 = 3600 ÷ 3 = 1,200, meaning that a volume of 1,200 gpm is needed for knockdown. Any involvement in the truss loft will require an additional 1,200 gpm; use large lines and large caliber streams. After knockdown, you may wye off 1 3/4-inch lines for overhaul.

For penetration and reach, opt for straight streams, either smooth or combination. Fog streams will do several things to a fire in a strip mall. A fog stream directed at nearby flames will push fire deeper into the occupancy and possibly up into the truss loft. Yes, steam will be produced, but the space is too large to contain the steam for proper cooling. Fog streams may also harm viable victims in the adjacent exposures. Finally, a fog stream may not allow you to reach the seat of the fire before it is converted to steam. This may hamper extinguishment.

How many members will it take to search this 60 x 60 pool supply shop in the 15 to 20 minute window?

You must define the term normal business hours by the types of occupancies present. A bar may be open until two or three in the morning, whereas a tax accountant will probably be closed by six in the evening. If the parking lot is totally empty, then search can be placed lower on your to-do list. Give more serious consideration to search in strip malls with residences upstairs.

Normal ventilation practices cease when dealing with a fire in a strip mall. The best-case scenario is one in which the fire vents itself at the roof. Unless the fire starts in that area, however, self-venting fires normally indicate advanced fires that have entered the truss loft. This means that flames will also be traveling horizontally through the truss loft. You'll need to concern yourself with exposure protection immediately, if not sooner. Normally, if the roof comes off one of these occupancies, everything below the roof will be destroyed as well. However, when the fire vents itself, suppression forces benefit from the effects of natural ventilation over the seat of the flames.

The second-best scenario is one in which the fire hasn't vented itself, but a quick knockdown immediately allow you to direct interior lines up into the truss loft. This should be done from doorways or arches near bearing walls. After a quick knockdown and flushing of the underside of the roof, get out of the building and let it set. Advance on the exposures and

A flat roofed Strip Mall with a large front façade. Collapse zones should be established if fire involves any portion of the façade.

A Strip Mall with a gable roof assembly. Trusses frame the form of the gable roof. Expect a huge lumber yard between the ceiling and the roof decking.

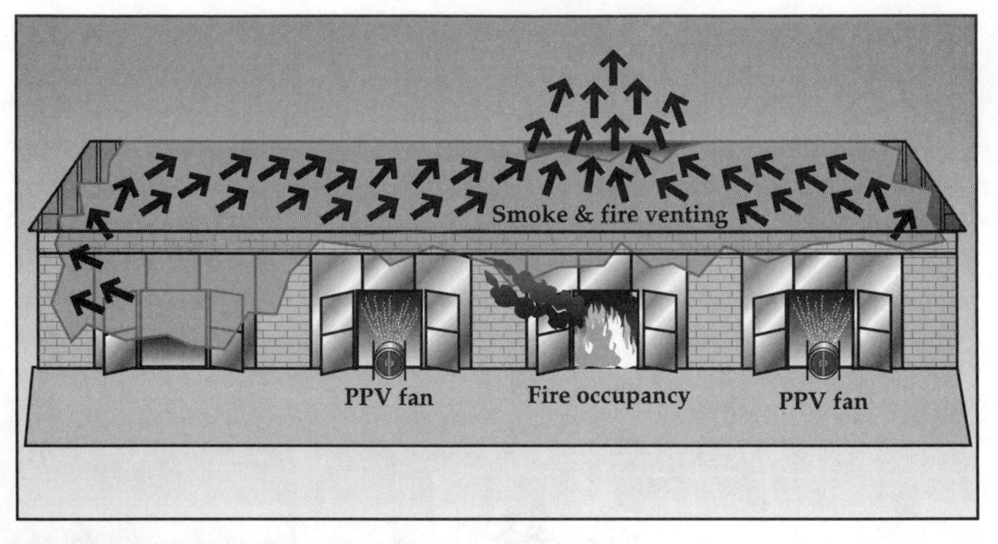

Utilization of PPV at a Strip Mall fire.

ensure that lateral spread has been stopped. After PPV has been running for a while, go in to check for interior extension.

I don't intend to debate the pros and cons of PPV, although I do have my concerns about it. I have listened to the debates at conferences and have even participated in some. I know how we use it in Toledo, and I also know how it's used in Columbia, South Carolina, as well as in New York City. New York and Columbia are at opposite ends of the PPV spectrum, and Toledo is in the middle.

I realize that PPV is a tool and that all tools have application in a particular context. A hammer is great at setting nails, but it's terrible at driving a screw. The concept of charging a truss loft with air to control fire spread back toward the area of involvement is new and needs more testing. In theory, pushing the products of combustion toward a vent hole will work, and I wouldn't hesitate to use it if conditions were right. Still, it's not a perfect science. Certainly, as fire departments all over the country accumulate experience with it, much more will be said on this tactic in the years to come. This is what is being considered, if a fire in a strip mall has vented itself at the roof upon arrival, initial actions are to get into exposed occupancies on each side and start PPV inside the exposures with the hopes of forcing air up and out of the vent hole.

Always consider placing backup lines in fires involving strip malls. The focus of the backup team remains the same from fire to fire, but the particulars of the deployment will certainly change. In a working fire in a strip mall, one specific backup crew should be assigned to the attack team. The backup line should be positioned approximately two-thirds of the way between the point of entrance and the attack nozzle. Once the line has been stretched, the

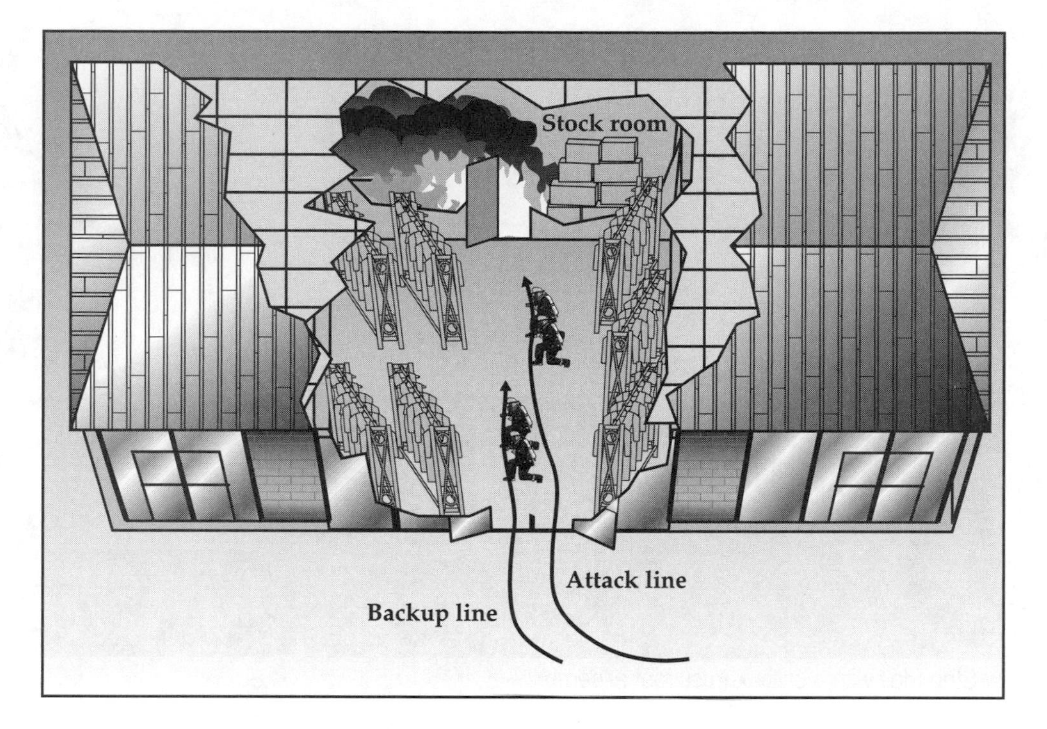

The positioning of a back up line at a Strip Mall fire.

officer assigned to backup should follow the attack line up to the attack officer and check on their safety and progress.

Because the fire loads in these occupancies can be huge, the backup officer should at least consider pulling a 2 1/2-inch line. If a fire breaks out in the truss loft, then a line of significant proportions may be needed to hold the flames in check as the crews back out. I know that I would feel better as an attack officer if the line backing me up had some punch to it. Some will say that it's difficult to pull and maneuver a 2 1/2-inch line, and there is some truth to this, but once the backup line has been deployed, it doesn't have to move much. Two firefighters can effectively manipulate it, suffering only short-term fatigue.

As always, the backup line should follow the path of the attack line. This ensures that the officer assigned to backup always knows the location of the attack crew, and it lessens the chance of having opposing lines. As crews become available to take lines into the exposures, you should prepare additional backups. One line per exposure will almost always be required for every exposure that has fire in it. Search crews can usually be shadowed by a single backup.

Checking for extension is done at the party wall dividing the two occupancies. The best thing that you can find in the ceiling is a firewall that penetrates to the roofline. The worst thing that you can find is a raging fire in the truss loft. If you can't stop the progression of the

A new Strip Mall with a saltbox truss roof assembly.

flames here, then move to the next exposure to make your stand. Pull the ceiling. Hopefully you'll be ahead of the fire. Report on the conditions to the incident commander and, unless directed otherwise, hold your ground. Maintain a vigil at the party wall, especially at the ceiling level. Don't leave unless you're told to do so or conditions warrant. If you're operating in smoke and need to leave for a bottle change, make sure that the IC is aware. If you believe that a replacement crew is necessary, inform the IC.

Pay close attention to the ceiling area, but also check the party wall occasionally for heat. Use your bare hands. Report any changes to the IC.

Unlike the roofs in garden apartments, the truss-roof assemblies in strip malls can collapse with disastrous results. In part, this is because the dividing walls are farther apart, and there are few cross or partition walls as there are in dwellings. The members of a rapid intervention team will need to evaluate the potential of a strip mall for collapse. They may choose to stage at the rear of the building, although opening up a doorway back there will likely have an adverse effect on PPV operations. Access is always available at the front of a strip mall, whether through the door or the display window. Also, don't overlook the possibility of breaching the party walls, which may or may not be reinforced. If you encounter rebar, use a hydraulic extrication tool to cut or spread it.

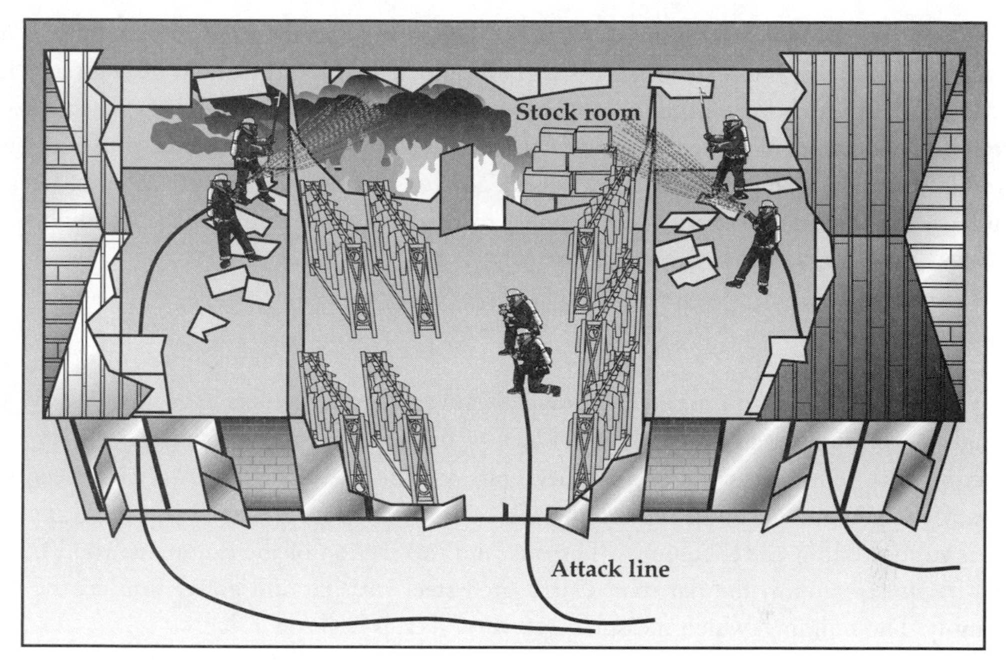

Crews pulling ceiling in exposures at a Strip Mall fire.

When checking for extension in this fire, crews must get into each exposure and pull ceilings along the wall abutting the fire occupancy. In this instance, Exposure "D" should be checked first if staffing is limited.

When it comes to overhaul in strip malls, there is one operant word to remember. That word is hydraulic. A fire-weakened truss is nothing that you want to stand beneath and tug on. Use a 1 3/4-inch line and don't get aggressive. Cut back on the pressures when hitting overhead, and stand in protected areas. Adjacent exposures are good when you can shoot through the open ceilings. Doorways are good. Master streams from outside are even safer. If you have any doubts, get a structural engineer to check on the stability of the building.

Scenario

One new strip mall on the edge of town always catches firefighters' eyes as they drive by, though it seemingly went up overnight. Three of its exterior walls are of true ordinary construction with reinforced concrete masonry units, steel lintels, and glass. A brick veneer wall hides the CMU surface up to the canopy. The canopy projects forward from the front wall and is supported by CMU columns. The face and top section of the canopy is wood frame. Steel-bar joists support the flat roof. Corrugated steel with tar and gravel finishes the roof assembly. The building, which measures 240 X 60 feet, is built on a slab.

The case study Strip Mall. This typical Strip Mall has a flat roof assembly.

Seven businesses occupy the mall. From left to right, they include a bistro, a shoe store, a dry-cleaning establishment, a men's clothing store, a tavern, a travel agency, and a store specializing in bedroom furniture. Each is separated from the next by a CMU party wall that rises twelve inches above the ceiling level, except the bistro, which has a party wall that extends to the corrugated steel deck of the roof assembly. This was done to satisfy code requirements that call for sprinklers for areas over 12,000 sq. ft. Due to this one firewall, none of the occupancies have sprinklers.

Along the back of the mall runs a ten-foot driving pad. Every thirty feet or so along the pad are dumpsters, set up tight against the building. The back wall has only one door per occupancy. Each is hinged on the inside. None of the doors have windows. There are no doorknobs, either–only keyholes.

The fire starts in the back section of the travel agency at 1930 hours on a Saturday evening. A bad switch on a coffeemaker is at fault. No one is in the unit at the time.

The response is triggered by several calls from the tavern and the furniture store, each reporting an odor of smoke. As the first-arriving engine maneuvers through the nearby intersection, the officer can see smoke rising from an indefinite source. He has been to a couple of dumpster fires at this address before, and he figures history may be repeating itself. Drawing nearer, he sees several people waving the engine toward the furniture store, the interior of which is hazy. The officer does a double take, however, and realizes that the seat of the fire is in the darkened unit next door. Moreover, smoke leaking out from under the canopy indicates that flames have already begun to travel in both directions through the truss loft.

Small-Department Response

The officer of the first-in engine must do two things. He must assume command, and he must take steps to flow as much water as possible, as quickly as possible. For a 20- by 60-foot unit that's completely involved, he'll need a deck gun. For fifty-percent involvement, he'll need a 2 1/2-inch line. He might get away with a 1 3/4-inch line, but he would have nothing left over for bad math or unforeseen problems.

Deck guns work best. Take out the front window and have at it. The next-in unit should either augment the water supply, pull additional lines for fire control, or check for extension in the threatened exposures. In the scenario described above, fire has already entered the truss loft and begun to travel. Given that the alarm was sounded by people in both the tavern and the furniture, exposures B and D, it's not unlikely that the final stand will have to be made from the clothing store, which is exposure B-1. Pull the ceiling in the exposed occupancy and wait for the fire to come to you. It won't hurt to flow water at the ceiling of the occupancy from which the fire is spreading. If the span is long, a 2 1/2-inch line may be necessary just to sweep the ceiling and the underside of the roof. Let the water flow, and don't stop until the IC tells you to do so.

The case study fire viewing conditions upon arrival.

With aerial streams from above and big lines from below, you can stop the progression of fire in most circumstances. Take away one of these elements, and your chances will diminish.

Normally, search shouldn't be considered a secondary assignment, but if you don't darken a fire in a strip mall quickly and check the exposures, the rapid spread of flames will likely nullify your best attempts at finding live victims. If manpower is extremely low, as in a small department, the task of search should probably be a secondary assignment given to second-alarm or mutual-aid crews. If possible, the police can search all occupancies devoid of smoke and ahead of the fire.

Overhaul will be a sizable task. Piles of burned stock can be extremely heavy when wet. If the fire is at night, it may be best to keep a crew on the scene to darken any flare-ups and wait until morning for fresh crews to handle the hot spots. Salvage will also be a chore.

Remember that the loss of a few businesses in a small community can be devastating to the local economy. Anything that you can do to help them will no doubt be appreciated. Still, when the fire department is constrained by limited resources, the secondary assignments will normally be assigned to second-alarm crews and mutual-aid units.

Medium-size Department Response

The first-arriving engine of a medium-size department has some distinct advantages over the first-in unit of a smaller department. More manpower means fewer split crews and more

The case study fire 15 minutes into the fire.

simultaneous operations. Still, the primary concerns and thrust of the incident remain the same: Knock down the fire and quickly check for extension.

As before, the second-in unit will have to augment the initial attack or hunt for extension in the most endangered occupancy. The third-in unit must move to the other exposure and look for extension there. Given the advantage of having more companies, a medium-size department may be able to initiate early search in the occupancies where the threat to life is severe but possibly survivable. There must be enough manpower to conduct the search appropriately, meaning safely and rapidly. Many of these occupancies have complicated floor plans. Parts of the interior may be loaded with stock. For search operations, backup lines and rapid intervention teams are mandatory.

Large-Department Response

The first and second engine should secure adequate water supplies and handlines to fight the fire. In well-advanced fires, the deck gun may be preferable. Two deck guns, shooting from engines spotted next to each other, can produce a tremendous amount of firepower.

The crew of the third-in engine will take a line into the most severely exposed occupancy and pull ceilings, either with pike poles or hydraulically. These members will then use the hose to cool down areas above the party wall.

The first-in truck will provide an outside vent man to take out windows in the front, if needed, and to work with the forcible entry team in the rear of the complex. If searches are required, the remainder of the first-in truck crew will begin this task. The second-in truck will probably set up in the direction of spread and cool the top surface of the roof. The remainder of this crew will assist with search operations.

Chapter Nine Questions

1. A __________________ is defined as a one or two story lightweight building that houses several different businesses under on roof.
2. ____________________ is a key concern in a working fire in a strip mall.
3. Generally, there is little problem with access to the rear of a stripmall. True or False
4. Life safety is a prime concern in strip mall fires. True or False
5. Built-in fire protection is generally sufficient in strip malls. True or False
6. The floor assembly of most strip malls is ______________.
7. The author describes the strip mall as a closed container with excessive fire loads and a ________________ for a lid.
8. After lines are in operation in a strip mall fire, the next action to take is to ____________.
9. What is the NFA's fire flow formula?
10. The author states that in his opinion, the best case scenario as it relates to venting a working fire in a strip mall is ______________________.
11. What route should the backup line take into a strip mall?
12. What are the two reasons for the answer to the above question?
13. Checking for extension at a strip mall fire is done at _______________.
14. What is the key concern to RIT at a strip mall fire?
15. What is the key word when discussing overhaul at a strip mall fire?

Questions for discussion

a) Compare and contrast today's strip malls to yesterday's taxpayers as it relates to construction.
b) Compare and contrast exposure protection efforts in a strip mall as opposed to single family residential structures.
c) Discuss the author's comments concerning "taking the glass" as it relates to quick knockdowns at strip mall fires.

Chapter Ten

Fighting Fires in Vacant Commercial Occupancies

Most of the occupancies to which firefighters respond are single-family residences. It would stand to reason, then, that the majority of firefighter casualties would occur at these incidents, but that is not the case. Proportionately, more firefighters die and are seriously injured in vacant commercial occupancies than in residences.

The term vacant commercial occupancy can mean different things to different people. Some will immediately think of an old heavy-timber warehouse with boarded-up windows and weeds growing all around the structure. Others will picture a decaying brick-and-mortar taxpayer in a run-down area of the city. Still others may envision nothing more than a vacant store in a local strip mall. No matter their exact type. As a group, vacant commercial occupancies constitute some of the most dangerous grounds on which to face the enemy.

Firefighter safety should top list of the incident commander's concerns, and the rules of engagement should constantly be on his mind. Two key factors are implied when describing a building as a vacant commercial occupancy. The first is that some or all of the stuff that used to be present in the building is gone. The second is that there should be no civilian life hazard involved.

Recalling the fundamental rules of engagement or risk assessment philosophy as it pertains to our industry, we are reminded that firefighters will take great risk to save life. In theory, however, the building isn't occupied, so that aspect of the policy shouldn't apply. We are reminded also that firefighters will take some risk to save property. If all of the valued items were hauled away by the former occupant, the bank, or the government, however, what can be the worth of anything left behind? That leaves us with the structure itself. Given an empty unit in a sprinkler-equipped building, we can see that the property probably has great value, and firefighters should take some risk to save as much of it as possible. The dollar value of a dilapidated taxpayer or some

A two story "ordinary" Vacant Commercial with an occupied exposure on side "D". *Credit Coleman*

old mill would certainly be more in doubt. The last part of our credo is that firefighters should take no risk for life or property that is already lost. This I interpret to mean property that has already been lost to fire or that has no value prior to the fire.

In point of fact, the first concern with a fire in a vacant commercial property is the risk to firefighters. To understand the risks, consider these buildings as falling into one of two categories: new vacant commercials and old vacant commercials. Besides being built more recently, those in the first category have some worth. Generally they're vacant because the former occupant went out of business and the building is waiting for another owner or occupant. Sometimes the building is awaiting its very first occupant. These buildings are usually up to code and have little structural damage. Moreover, there tends to be little rubbish or leftover stock in them.

Older vacant buildings are a different story. Most of them were remodeled during the long years that they were in use. Some of the additions may not have been put in by professionals or ever approved by local building authorities. Some may have compromised the integrity of the structure, such as rooftop air-conditioning units. Many of the old properties don't have equipment for automatic fire protection or detection. Pipes, valves, and other essential components may have been stripped out of the structure by scavengers and vandals. Holes and weakened surfaces may be common. Vacant commercial properties usually aren't

Collapse should always be considered in fires in Vacant Commercial Occupancies. Knowledge of building construction will aid in this decision making process.

A new Vacant Commercial in the downtown district. This building is under renovation in the Theater district of a mid-sized city. *Credit Coleman*

This "typical" Vacant Commercial Occupancy once housed a harmonica factory. *Credit Coleman*

up to code. They can have substantial damage and be littered with rubbish and structural trash. In short, they can be very dangerous.

Getting into a vacant commercial will either be difficult or easy. Either they're boarded up tight or they're wide open. Both circumstances present dangers. Wide-open ones are more likely to have holes in the floors, missing railings, missing stair treads, and other damage. Scavengers will take almost anything.The site may also be a dumping ground for hazardous materials and other refuse. Expect to find asbestos. Keep overhaul to a minimum, and only perform it under good lighting and with a safety officer assigned strictly to monitor this operation. On the positive side, open buildings pose little or no danger of backdraft, as do those that are boarded up. Closed environments may also experience early flashover. The interior of a tight building will be dark, and lighting must be set up early after extinguishment, prior to overhaul.

The question of construction type, a parameter for any building, is also a major consideration for vacant occupancies, and perhaps more so, depending on the condition and age of the structure. Since load-bearing structural members may be missing, expect early collapse. Have your dispatch center give frequent and regular time checks, such as every ten minutes. Assign one safety officer to monitor the building for signs of collapse, and pull everyone out

Exposure protection is a key consideration in initial line placement at a fire of this magnitude involving a Vacant Commercial Occupancy.

A vacant apartment building under renovation. *Credit Coleman*

This Vacant Commercial is tightly sealed. A very hot fire and perhaps even a back-draft could be produced under these conditions. *Credit Coleman*

A typical Vacant Commercial Occupancy located in the inner-city. *Credit Coleman*

when you feel it's time to do so. Don't let crews keep going at it when you feel that the structure is nearing its limit. If an entire floor is involved in a building of ordinary construction, expect it to last for about ten to twenty minutes before collapse. And that's a good building, with all or most of its load-bearing members intact. A building that's vacant will probably have missing components and the by-products of shoddy renovations. When collapse does come, the walls may go in or out. Establish collapse zones early.

A fire in a vacant commercial occupancy will require a lot of restraint on the part of the first-arriving officer. Once he has taken into account the type of construction, the value of the structure, the location of the fire, the presence of exposures, the weather, the risk-to-benefit ratio, and other such factors, he may commence operations. Some of the most exciting and some of the most boring fires that I have ever been to took place in vacant commercial occupancies.

Fires in these buildings can be large in scope. Since there are theoretically no occupants around, the sounding of the alarm will likely be delayed. The building may be heavily involved by the time the first unit arrives. Old oil-soaked floors may contribute to the spread of fire. Holes in the walls and other surfaces may foster widespread extension. In Toledo we are considering a defensive-only stance, based on a rating system of known buildings, prior to the alarm. Other communities are doing the same. Vacant occupancies that exhibit structural decay or those that have had previous fires are dispatched as a Code Red, meaning that firefighters won't enter the structure if it's involved.

This is tantamount to a let-it-burn policy. Vacant occupancies that have worth, are structurally sound, and that have never experienced a fire are dispatched as a Code Yellow, meaning that cautious interior operations may be conducted. Code Green is for all other buildings. These ratings are provided by company officers and are recorded on the computer-aided dispatch system back in the alarm office.

At bottom, risk analysis determines whether an offensive or a defensive approach is in order. Given the aggressive nature of an offensive attack, you can expect a faster knockdown time. Conversely, you should anticipate a slower knockdown time from defensive operations. Offensive operations are clearly be more dangerous. Even so, hundreds of firefighters are injured each year while in the defensive mode, and the restraint and coordination with which you mount these operations offer no guarantee of safety.

Offensive attacks should be mounted from the unburned portion of the structure, if possible. This will help complete the first part and most important part of the extinguishment process: confining the fire. When going in from the unburned portion, it may take a little longer to get the line in place, but it will also effectively check the fire at the area of first water.

There aren't many reasons not to pull a 2 1/2-inch line for the initial attack in any vacant commercial occupancy, and manpower constraints certainly don't constitute a reason. Two firefighters can stretch and handle a 2 1/2-inch line with little problem. With a smoothbore nozzle and moderate nozzle pressure, you'll get a lot of reach, penetration, and knockdown power. More than 300gpm can be thrown by a single member on the nozzle. If time and the fire permit, get the line in place before you charge it. Once you've achieved knockdown, you

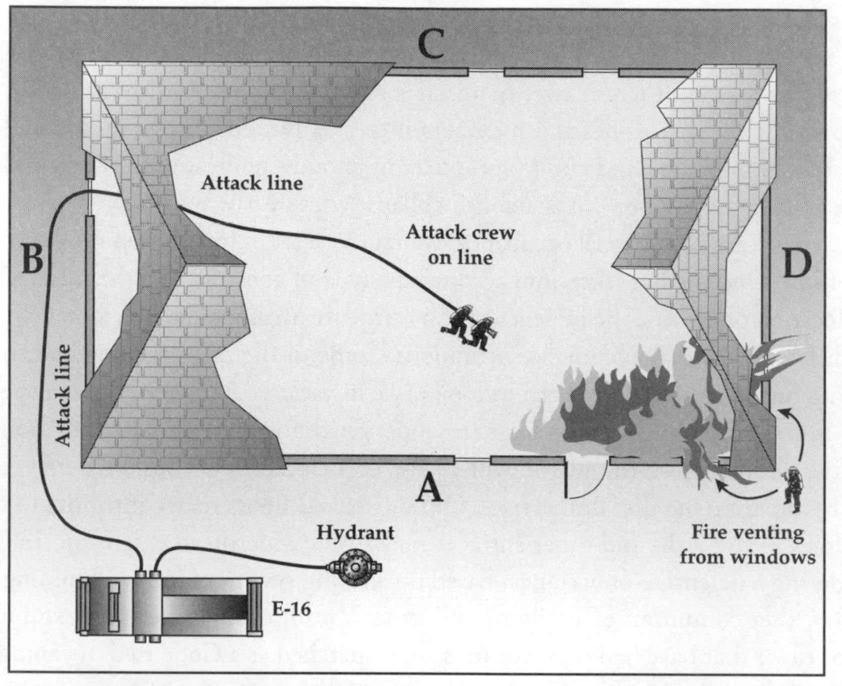

Proper line placement at a working fire in a Vacant Commercial. The line is taken in with the thought of confining the fire to the smallest area possible. If the line were taken in through the front door, the fire may have been spread needlessly.

In offensive operations, initial line placement is the first leg in the extinguishment process. (Confine the fire to the smallest area possible.) In this instance, a line in the front door near side "D" will work well.

Roof assemblies need to be pre-planned. This Vacant Commercial contains penthouses which may prove beneficial in ventilation efforts. *Credit Coleman*

can split off smaller mop-up lines from the 2 1/2-incher. As a rule of thumb, if you can't knock down the flames with a single 2 1/2-inch line inside, you might consider going defensive for a while.

My preference is to use an indirect attack off the ceiling. This will provide reach and great knockdown power with the least disruption of the thermal balance. Additionally, it helps to control high temperatures at the level of the ceiling. The exception to this may be in a vacant motel or hotel with tight hallways. A combination nozzle on straight stream will provide the added feature of a fog stream to hold fire if it advances down a hallway. Breakaway nozzles that are smooth bore and have a combination nozzle attachment provide the best of both worlds, as long as the officer has the combination nozzle in his pocket and not back at the engine.

Always use backup lines when conducting offensive operations in a vacant building. Again, the 2 1/2-inch hose is the line of choice, and it should be the second item taken in through the point of entry. As in other situations, the backup line should follow the same route as the attack line, and it only has one purpose. This means that it should stay approximately one hose length behind the attack crew to protect their path of escape. Once the fire has been knocked down and the IC gives approval, the backup line can be used for other purposes, such as extension or overhaul.

At some fires, initial defensive operations are warranted. This would be one such example.

This vacant apartment building might be rated "Code Red". It has had previous fires and some structural damage. *Credit Coleman*

This vacant store-front might be rated "Code Yellow". Operations under due caution would be indicated. *Credit Coleman*

This vacant store-front with apartments overhead could be rated as "Code Green". *Credit Coleman*

Defensive operations are holding operations, and the initial placement of lines is no less important here than it is when operating in the offensive mode. Position them from the unburned portion of the structure, if possible, so as to confine the fire to the smallest area possible. The fire will normally be working toward areas where oxygen is abundant, driven by wind and convection currents. If the attack is made from the burned area, then you'll only facilitate this progression. If attempts to get ahead of the fire prove to be impossible or impractical, then attempt to flank it as much as possible. The key is first to get ahead of the fire and stop its progression.

Elevated streams from master appliances should be the tools of choice. Quick knockdown is the objective. Deck guns, monitor nozzles, and 2 1/2-inch handlines will all work, given that the objective is simply to flow enough water to effect extinguishment. There's nothing wrong with knocking down a fire from the outside of a vacant commercial property, then going in to mop up. Knock it down, kick on the lights, and turn on the fans. Survey the building for structural problems. If everything looks okay, then go on in to finish off the fire.

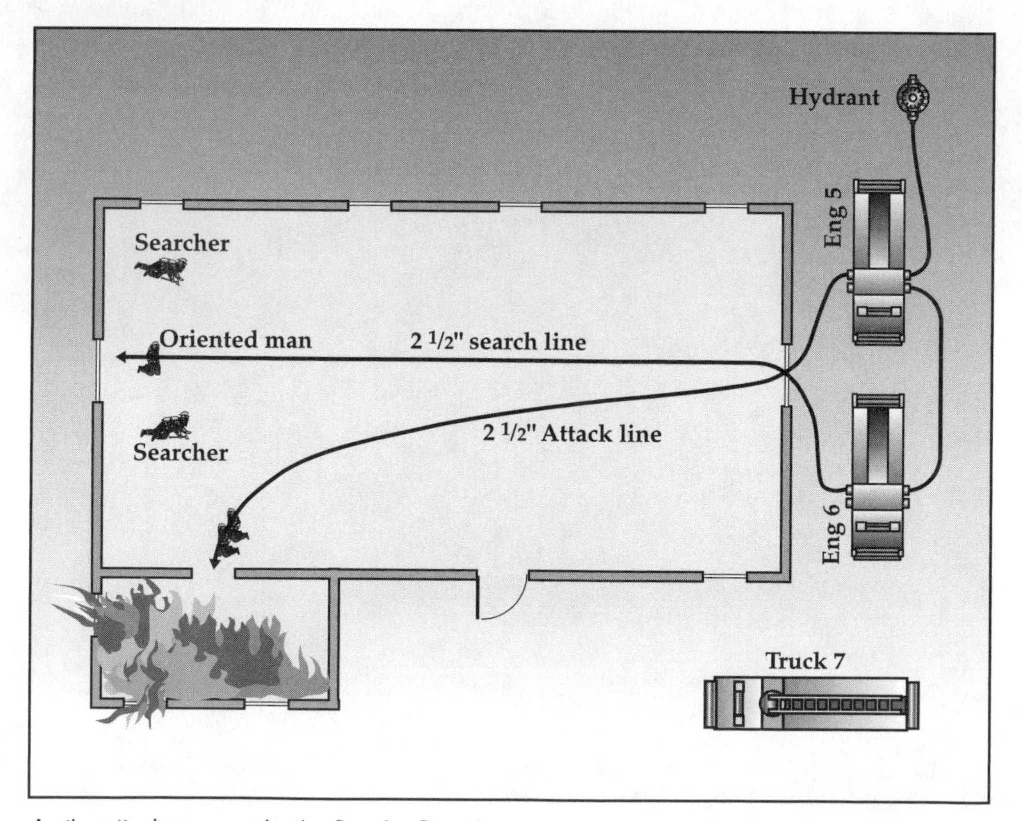

As the attack crew works the fire, the Search crew stretches a 2-1/2" line to search off of.

Search will be lower on the incident commander's to-do list at a fire in a vacant commercial occupancy; however, you should still conduct primary searches if you believe that there might be savable victims inside. There could be children playing inside, or homeless people. There could be police officers that arrived first on the scene and got caught while making a quick sweep of the building. Look for open windows, doors, and other entrance paths. If the building appears tight, then put search lower on your list.

If search is warranted, use caution. An oriented search, working off of a charged 2 1/2-inch line, offers the best protection. The oriented man remains on the line while the searchers work off of it, much as in the team search method. The differences between these two methods are that the oriented man stays on the line and doesn't search, and the team has a 2 1/2-inch hose rather than a length of 3/8-inch kernmantle rope. You'll know immediately if something horrendous happens to a hoseline—it'll go limp! You might not know whether a rope has been compromised until it's too late.

Thermal-imaging cameras are extremely valuable in these occupancies. If only one camera is available, then the oriented man should have it, since he has time to survey the scene around him, rather than just be focused on search.

Ventilation is essential if you intend to save the structure. Many of these buildings need little ventilation effort. Missing windows and leaky roofs may make any coordinated ventilation efforts futile. In such instances, when the building has already been opened up to the

This roof, with a monitor skylight, should require little effort to initially vent. Take the glass in the sky light.

point that the pathways for smoke and heat would be unpredictable, then little or no effort should be expended on ventilation. Let the wind and the convection currents do their thing.

If ventilation is to be attempted and the building is somewhat tight, then choose your weapon! Mechanical ventilation using fans will probably be the least-chosen method in large occupancies, given the difficulties and uncertainties associated with moving volumes of air through cavernous spaces. Natural ventilation will probably be the best method to use in a vacant commercial occupancy. Natural horizontal venting through windows and doors may be the least effective, but it'll likely be the safest. Take windows from the lee side first, and take them from the highest point in the window frame. Then take out windows on the windward side. Take these windows completely out. If this is the only safe means of venting, then you're pretty much done.

If you're going to open the roof, then your next concern is the type of roof on the building. Cut wooden gable roofs at the highest point possible, at the ridgeboard. Carbide-tipped chain saws will probably be the fastest on commercial roofs. The amount of fire and its proximity to the upper levels of the building, as well as time, manpower, and tools will all dictate the size of the hole to render; the bigger the better. A ten-foot by ten-foot hole is very ade-

This is an example of a tight vacant commercial building. Again, this type of Vacant can produce hot fires. *Credit Coleman*

A working multiple alarm fire in a Vacant Commercial Occupancy. The primary concern at this fire is safety to crews and confinement. *Credit Ron Jeffers*

quate; if constraints only allow for a five-by-five hole, then that's okay, too. Pretty much the same rules apply for flat roofs. Skylights, if available, make for the safest, easiest method of venting a flat roof. Warn the crews below, take out the glass, and get off the roof! If holes are required, cut them and get off! These are vacant commercial occupancies. For the most part, they're trash buildings. Neatness means nothing here. Carbide-tipped chain saws and K-12 saws are the fastest. Axes will work, but they're labor-intensive and certainly not as fast.

Remember that venting is done for three reasons: (1) to channel smoke and heat away from occupants, (2) to reduce damage from fire and smoke, and (3) to make our jobs easier. For the most part, reasons one and two aren't important for this type of occupancy. If we're committed to a defensive operation, then even the third reason may not apply. Ordinarily, I like to have the vent team stay on the roof after the hole has been cut, but at fires involving vacant commercial occupancies, I prefer to get them back down on the ground as soon as possible.

Checking for extension can be a tricky assignment in a vacant commercial, the main problem being the integrity of the structure. An incident commander normally has three choices with regard to this operation. First, he can assign a second crew to check for extension concurrent with attack operations. This is dangerous but at times essential so as to check for probable spread. The second option is to reassign the attack crew to extension as soon as the main body of fire has been knocked down. This is the normal course of events when staffing is low. The third option is not to assign a crew at all, as when it would be too dangerous to send a crew above the fire at any time throughout the incident.

The process of checking for extension is normally carried out in three stages: above the fire, around the periphery of the fire, and below the fire. The type of structure and the location of the flames will determine the types of tools that will be needed for overhaul. Fires in buildings of concrete, of course, will require little checking for extension, other than pipe chases and other customary routes. Normally, when such areas are located, a flushing with a hose stream will usually suffice. Rarely will you need to open walls in a concrete structure.

Fires in buildings of ordinary construction may require more work. Generally, the only attention mandated for the exterior walls will be around the window frames and other structural openings, but the interior walls can be of a variety of materials. If the partition walls are also of ordinary construction, little opening up will be required except, again, around the various utility channels. The ceilings and floors will normally be of wood, and these may require full overhaul.

Newer buildings, predominantly of wood-frame or ordinary construction, will have truss construction in their floors and ceilings. Such occupancies normally have drywall on steel or wood studs. Checking for extension in these buildings will be similar to the same operation in newer residential occupancies.

Exposures can play a key role in these fires. Unlike most incidents, the exposures become the prime concern in many of these emergencies. Prioritizing which to protect first is often a judgment call. Some of the factors to consider are the wind, the distance from the source fire, the life safety issues in any exposure, the value of the buildings, and the historical or religious value of the structures. When the source fire is in a worthless building, then the exposures take on additional value. To protect them to the fullest, play streams from large-caliber lines onto the exposed surfaces to keep them below their ignition temperature. If the distance is close and the level of heat is high, maintain the exposure line until the source fire either dies

What would Command's concerns be at this fire in a vacant warehouse?

down or has been extinguished by the attack crew. If the source fire is smaller or the distances are greater, lessening the risk of extension, then the exposure line can be played between exposure protection and extinguishment. This is an option, but if the exposed wall begins to heat up and give off vapors, then maintain the line on the exposure only.

Once a line to the exposure is in place, then the remainder of the crew or another crew must get inside the exposure to ensure that fire doesn't get into the building. They must feel the walls on the exposed sides and remove any flammable materials from around the windows. They should also continuously monitor the window frames and attic areas.

I believe that if any sort of building defines the role of a RIT, it's the vacant commercial occupancy. Proportionately more firefighters are killed and injured in these buildings than in any other type. Construction features, access routes, the location of the fire, and the number of personnel inside will top the list of the RIT officer's concerns. In determining the best means of access, the members of the RIT should be looking for balconies, fire escapes, add-on stairs, and other features that can provide a way to the upper floors. In Toledo, we require the RIT to inform the incident commander if a ground ladder is raised to the rear of the building. If a ladder will facilitate any rescues, the RIT should ask the incident commander for permission to raise one. From the time of the 360-degree walkaround and onward, the members of the RIT should be evaluating fire conditions. Their observations should be reported to the command post and weighed against all other incoming information.

A fire in a vacant commercial occupancy is one of the instances in which multiple safety officers may be warranted. In doing this, start with an interior and an exterior safety sector. Due to the two-in, two-out rule, any interior safety officer should have a buddy. The exterior safety officer may operate alone. If crews and the fire are on the upper floors, it's vital to establish a second means of egress as soon as possible. Ground ladders set up by the RIT will normally serve this purpose, but tower ladders and other aerials will also work well if they're available.

Scenario

The fire occurs in an old metals manufacturing company constructed in the mid-1930s. The main building is four stories tall, but a section on Side D only rises to three. It is comprised of ordinary construction and has steel columns as additional support. Each of the lower three floors is approximately 15,000 square feet in area, and the fourth floor covers about half that. The floors themselves consist of wooden planks, which have been soaked over the years with cutting oils and other volatile liquids. They are supported by wooden joists set into pockets in the walls. The roof is flat, made of planks, covered with tar, and supported by wooden rafters. There is a fire escape on Side C of the building. For the most part, the windows are intact, and some have security bars. There's an occupied apartment building along Side C, creating a severe exposure problem.

The building does have automatic sprinklers, but vandals and scavengers have compromised the system to the point of uselessness. There is an open elevator shaft in the building, and some of the stair risers have been removed from both open stairways, one on Side B and the other on Side D.

This building is the case study building. If you were RIT, what would your concerns on side "C" be? *Credit Coleman*

The front of the scenario fire building. *Credit Coleman*

At approximately 08:30 on a Sunday morning, an arsonist ignites a trail of gasoline inside the building, beginning at the stairway on Side B, looping around the elevator shaft, and terminating at an open window on Side D, his point of exit. The first officer on the scene finds a heavy volume of fire inside a structure that he has considered for years to be a sitting duck. His first and foremost thought is the occupied apartment building along Side C. His strategy centers on saving that building, since that's where the life hazard is, and the fire building itself clearly isn't worth the trouble.

Small-Department Response

Mounting an interior attack would take at least all of the officer's first- and second-alarm units. Even if he commits to an interior attack, he might not have enough manpower to beat back the fire. He therefore opts for a holding attack, designed to keep the fire to the smallest area possible. The flames will probably spread anyway, but his hope is at least to confine them to the original building and not involve the exposure. The officer wants to surround the building with aerial devices and master streams. He knows that the best way to apply them is first to cover the areas where you don't want the fire to go. Another way of saying this is to get lines and water between the fire and what you want to protect.

In this case study, the fire was started by an arsonist in a vacant commercial building. This is side "A" and "B" of the fire building. The exposure in on side "C".

Given the initial shortage of manpower, the officer has a policeman begin the evacuation of the apartment building while his on-scene crews stretch large-diameter hose. No booster lines for a fire of this magnitude, even though the original building will be sacrificed. The game is to control the fire. Once the roof has self-vented, the fire will quickly consume the rest of the structure, eating along the most convenient routes of travel. Extinguishment operations will consist of darkening down the hot spots. The second-alarm units will handle the mop-up. Large fires such as this tend to become very boring. After the thrill is gone and the fire has been darkened down, monitor nozzles, aerials, and a few handlines will be able to handle the mop-up phase.

It makes no sense to allow personnel to enter a demolished building to extinguish hot spots. Hopefully we're more sophisticated than that nowadays. The safest way to handle the hot spots is to bring in heavy equipment to dig them out. Most mutual-aid companies can and should go back to their own jurisdiction at this point. A heavy-equipment mop-up is the province of the local department. A crane and a 1 3/4-inch line should be enough to handle the ruins.

Medium-size Department

With additional manpower responding to this fire, you might think that an interior attack would be feasible. Some officers of larger departments might believe that they can

This is side "D" and "C" of the fire building. The exposed apartment building is on the right.

muster the resources to attack this fire head-on. Older officers who have been bitten may prefer to lay out the strategy on their own terms, not the fire's.

The officer of a medium-sized department will have the resources on the first alarm to set up defenses in a shorter amount of time, as well as to assign a crew to the exposure building. Except for the first-in engine, which deploys a handline, succeeding units of the first, second, and third alarms should set up on all corners with master streams to control the spread, protect the exposures, and control the burn. As ever at fires in vacant commercial occupancies, the establishment of a RIT is a must.

Mop-up for a medium-sized department can best be handled by new companies. The initial units dispatched to these fires normally take a pretty good whipping. To best avoid fatigue-induced injuries, use ancillary companies or late arrivers and set up rehab. If you don't have enough staffing to accomplish this, then rotate your crews often. Follow the first-in, first-out rule. Even if you do have the manpower, there's no real reason to allow crews to enter this sort of building for mop-up. Again, have them watch for flare-ups and cool down the hot spots. Call in the heavy equipment.

Large-Department Response

The members of the first two engines pull big lines into the building. They can flow over 600gpm and handle a fire almost 2,000 square feet in size. The intention is to go in fast with plenty of water and quench the fire. The outside vent man opens up in front of the fire so that the attack teams can safely advance on it. With four or five men on each line, they'll have the maneuverability they need to advance.

The first-in truck provides the outside ventilation and forcible entry. It sets up in front of the building and provides ventilation on the upper floors as well. The remainder of the crew enters to search for life and fire. The third-in engine takes a line to the rear of the building and tries to advance on the fire. These members also have water in the rear if the fire gets ahead of them.

Additional crews are sent above the fire. The squad and the second truck will both take this position. With personnel throughout the building, construction becomes a bigger factor, and the IC asks for ten-minute time checks. If the crews aren't able to advance on the fire within any of these intervals, they'll be withdrawn. If being unable to advance on a fire in a vacant building doesn't make sense in the modern context, get them out and establish defensive operations.

The first-in truck of the second alarm should be used to establish rapid intervention teams. For the secondary assignments, rotate fresh crews into the ruins. Most large departments have plenty of personnel available to relieve the initial teams. And just because you're big doesn't mean you're stupid. Bring in heavy equipment to control the hot spots.

Chapter Ten Questions

1. A new unoccupied strip mall occupancy should be considered a vacant commercial occupancy. True or False
2. ________________ should top the list of the Incident Commander's concerns at a fire in a vacant commercial occupancy.
3. A department's risk policy should have no relevance as it relates to a fire in a vacant commercial occupancy. True or False
4. As it relates to access to vacant commercial occupancies, they are usually ___________ or ________________.
5. You can anticipate collapse in a vacant commercial of ordinary construction if an entire floor is totally involved for ________________ minutes.
6. Offensive attacks should be mounted from the ________________________.
7. The author's choice of offensive attack in a vacant commercial is an ________________.
8. A __________ should be the line of choice for backup in a vacant commercial fire.
9. With defensive operations at a vacant commercial, ____________________ should be the tool of choice.
10. What type of search does the author suggest at a vacant commercial fire?
11. The author considers a ____________ hole adequate as the initial ventilation hole in the roof of a vacant commercial.
12. If available, what is the safest and easiest method of venting a roof at a vacant commercial?
13. Once the exposure line is in place, the next action in exposure protection is to ________________.
14. What is the key concern to the RIT officer in a fire in a vacant commercial occupancy?
15. If using more than one safety officer at a fire in a vacant commercial, what is the author's suggestion on where they should operate?

Questions for discussion

a) Compare and contrast Risk Policy or the Rules of Engagement for fires in old as opposed to new vacant commercial occupancies.
b) Discuss the search of a vacant commercial building. What are the moral and ethical concerns in not commencing a search in an old vacant commercial?
c) The author gives some factors to consider in exposure protection in vacant commercials. Is your list consistent with the author and if not, what are the differences?

Chapter Eleven

Fighting Fires in Institutional Occupancies

The term *institutional occupancy* covers a range of facilities in which large numbers of people may be residing under somewhat confining circumstances. Hospitals, penal institutions, and schools are the quintessential institutional occupancies, since their construction features and firefighting concerns are all quite similar. Although nursing homes qualify as well, I am devoting an entire chapter to nursing home fires later in this text and so shall not specifically refer to them here.

The key problem that arises when fire strikes an institution is one of rescue, since many of the residents won't be able to rescue themselves. This places terrific demands on responding crews. To help stave off many of the difficulties associated with rescue, the primary strategic thrust becomes one of defend in place. The fundamental premise behind implementing such a strategy is that there are significant numbers of people to be taken out. Some hospitals house thousands of patients. The same holds true for prisons. The second given in the equation is that group rescue is rarely a possibility. There are times in apartment fires when a single firefighter can lead several civilians to safety, but this isn't possible when the residents are nonambulatory or present a security concern. Try to imagine struggling with a convict over your air supply. To be sure, there are minimum-security prisons in which a one-on-one rescue surely would be feasible; still, you must consider the number of rescues to be made. In hospitals, many patients can't even be taken out of their rooms unless they are to be disconnected from their life support and possibly die.

To save as many victims in an institutional occupancy as possible, defending in place becomes the premier option. The term means just what it says. Public responders defend the occupants where they are. Instead of removing them to safe areas, firefighters commit all of their efforts toward extinguishment and ventilation efforts. As tough-minded as this strategy

A modern high-rise hospital. Defend-in-place operations will probably be the strategy of choice. *Credit Coleman*

A fire in a hospital such as this would indicate a "defend-in-place" strategy.

may seem, operations conducted this way often save more lives than those in which responders concentrate solely on evacuation.

As always, the type of construction plays a significant role. Most modern institutional occupancies are protected with automatic sprinkler systems and standpipes. Older occupancies may lack sprinklers, but these buildings were normally built to last a long time. They were constructed of fire-resistant materials, and their components are normally massive, which translates into additional fire-resistance. These occupancies also tend to be very compartmentalized. A series of small cubicles for individual or joint occupancy in prisons and hospitals is the norm. The days of hospital wards lined with twenty or more beds are long gone in this country.

In older penal institutions and hospitals, the exterior load-bearing walls are predominantly of ordinary construction, sometimes in conjunction with steel columns. These buildings may be several stories tall but aren't what we would call high-rises. Four or five floors are usually the maximum. The interiors are supported by load-bearing walls and columns. The columns may be of steel encased in some sort of masonry; otherwise, they might be of solid ordinary construction. These buildings are sound, and if collapse occurs, it is usually precipitated by a fault in the floors or roof.

Older schools that are still in use today are predominantly of ordinary construction. These are generally more than one story tall but seldom exceed four. The interiors are supported by masonry or steel encased in masonry. Naturally, classrooms are larger than hospital rooms, but the structural integrity of a school is nonetheless improved by the compartmentalization of its interior.

A different sort of construction is present in newer penal institutions, hospitals, and schools. Modern penal institutions contain more steel and concrete, and they have less brick. Some of them have exterior brick veneer walls so as to blend into the neighborhood better, but steel columns will be the main support, and CMUs will span the area between the columns to keep out the elements. Modern hospitals are built of steel and concrete, and schools can be a combination of these materials; otherwise, they can simply consist of all-steel walls for both support and the exterior covering.

By the NFPA ratings, most or all of these structures would either be Class I or Class II structures, but don't be fooled by the ratings. Interior furnishings can produce very hot fires and deadly smoke in seconds. Since compartmentalization is the norm in hospitals and prisons, fires in these occupancies will likely be held in check by both the sprinkler system and the interior walls. Expect to find room-and-contents fires spawned in plastic mattresses and other furnishings. Rarely will these fires spread to other areas or rooms.

In older institutional occupancies, the roofs generally consist of wood planks with either tar and gravel or tarpaper weatherproofing. Most of these roofs are flat, supported by columns, wooden girders, and joists. You'll often find penthouses on these buildings that house stairwells or elevator equipment. Some occupancies may have skylights. Expect difficulty in venting a roof on one of these buildings. There may be layers of old outer weather-

A three-story nursing home. Each floor could house over 100 ambulatory or non-ambulatory residents. *Credit Coleman*

A modern one-story public school. Wings and long hallways will be the norm with this type of occupancy. *Credit Coleman*

proofing, and true two-inch wood planking isn't uncommon. Carbide-tipped chain saws and K-12 saws work well in these situations. Using an ax will consume both time and energy.

Newer roof assemblies consist of lightweight trusses, either of wood or steel bar joist. The profile of these roofs can vary and even display different characteristics in the same building. Whether the roof is flat or pitched, expect lightweight support for the roofs in these occupancies. The exterior coverings can range from metal deck to membrane to built-up to steel. The essential point of the matter is to watch the buildings go up in your jurisdiction and know them from the inside out.

Older penal institutions and hospitals have wooden floors, generally of two-inch planks, possibly overlaid with plywood. For sanitary reasons, most of these floors are covered with tile or vinyl. Support for the floor is provided by joists of 2 X 10 inches or larger. Such floors aren't susceptible to rapid, universal collapse. Rather, they give plenty of warning, generally in the form of profound sagging, sponginess, and attention-getting moans and groans.

Older schools have either old hardwood floors or poured-concrete floors covered with tile. The hardwood variety can have layer on layer of wax, plus later coats of ethylene varnish. Flame spread can be a problem with these floors. Large wooden joists support these floors and, again, sudden collapse isn't the norm.

This new low-rise prison presents several concerns. Site access, window access, interior obstructions and the possibility of an unruly victim population are among a few concerns. *Credit Coleman*

Newer institutional occupancies are of Class I or II construction, with floor assemblies of concrete, steel, or both in combination. Expect to find Q decking supported by steel joists, overlaid with linoleum or tile. Many of these newer occupancies will have a low profile and fewer than three floors. Some hospitals may be taller, but built-in fire protection systems and fire-rated compartments make them less prone to extension. Moreover, fires are usually detected earlier in them. Newer institutional occupancies tend to be very compartmentalized. Flame spread is usually held to a room or part of a floor by self-closing doors. By code, these structures must be protected by automatic sprinklers and have standpipe systems throughout.

In older institutional occupancies, the spread of fire is assisted by numerous openings and voids. Most of these buildings had wide-open, unprotected stairwells. Older schools, if involved, can produce an abundance of dense smoke. Varnished floors, lath-and-plaster interior walls, and wooden desks can all contribute to the generation of high levels of heat.

Older hospitals can burn quite well. Some of their wards were remodeled without a lot of thought given to the behavior of fire. If construction codes existed, they were often ignored. At times, the balance of concern came down on the side of profit rather than for the safety of the occupants. If any older hospitals are still in use in your community, it's essential that you visit and plan for them so that you'll know what you're getting into in the event of a fire.

Older prisons might also burn very well. They weren't designed to afford the creature comforts granted to residents of other types of buildings. Fire protection was never a prime concern during the planning of these occupancies. Prisons were built to be massive, and they could resist the effects of fire, but flames could still ravage the interior portions and travel extensively within. Although newer versions may even include some lightweight construction, they are also bound by code to meet modern fire standards.

The life hazard can be overwhelming at a working fire in an institutional occupancy. Some of the worst loss-of-life fires in America have occurred in hospitals, prisons, and schools. For a responder on the first alarm, the ratio of victims to rescuers can be greater than fifty-to-one. This alone should cause us to rethink our immediate concerns. It's vital that an incident commander know two things: first, whether the building is protected by automatic sprinklers and standpipes, and second, the capabilities of the crews available to him. Most departments have neither the manpower nor the training to accomplish multiple tasks. Given the potential scale of these sorts of incidents if operations go awry, it's best to concentrate on managing the cause of the overall problem–which is the fire.

Consider the math. If an IC were to forego a defend-in-place operation and decided instead to evacuate a hospital wing containing forty patients, many of whom are no doubt nonambulatory, he would need up to eighty firefighters to accomplish that task, with still no attention given to the fire itself. Some of these victims may be hooked up to life-support systems that, if not removed without care and expertise, could jeopardize the welfare of the patients. In an environment filled with smoke, it may be next to impossible to remove all of the paraphernalia supporting life and remove it correctly.

The interior hallway of an older school. *Credit Coleman*

One obstacle that must be overcome at fire in a prison or jail is access to specific areas. Bars can also present maze-like floor patterns in "zero" visibility conditions. *Credit Coleman*

To defend in place means that all crews must be committed to quick extinguishment and ventilation of the contaminated areas. Aggressive attack is the key. Multiple attack lines taken in from the same direction will help foster the safest operations. If standpipes are present, opt to use them over lengthy stretches, since doing so will help you avoid unnecessary lays and delays. Two lines are better than one. A fire stream will put out all the fire that it can reach in thirty to sixty seconds. In some of these occupancies, you may not have enough time to bring in more hose if the fire hasn't already darkened down within a minute. It's better to have more firepower than you'll need at the outset than to wish for it later. Don't waste time looking for more water after the initial attack. This is a prime opportunity to pull and work a 2-1/2-inch hose off a standpipe, as long as the connection is also 2 1/2-inches in diameter. If you connect with a 3-inch short section, you could split that off to 2 1/2-inch and 1 3/4-inch attack lines.

The placement of the first line will make or break the entire operation, but there will be few occasions in which the first stream shouldn't go between the most savable victims and the fire. For the greatest good, protecting the most, we must cut the spread of fire, heat, and smoke. Sometimes a small contingent of occupants may temporarily have to be placed in greater harm so as to do the best for the majority. Obviously, these decisions aren't easy ones to make.

Suppose a fire occurs in the front wing of a nursing home and there's access from both the front and rear of the wing. There are twelve rooms on each side of the hall, for a total of twenty-four rooms in the wing. The fire is in the second room from the front of the building. If left unchecked, it may spread equally in both directions. In such a scenario, the lines should be taken in from the rear and worked up to the flames, since this will allow you to protect the most victims with the least effort. Taking the lines in from the front might allow you to reach the fire a few seconds earlier, but the fire would almost certainly be pushed toward the main body of the wing. Any victims attempting to flee toward the rear entrance would be chased by billows of steam and smoke.

For proper control, multiple attack lines are essential. Once hoselines are in place and operating, ventilation should commence as soon as possible. For those who espouse positive pressure, fans can be placed and turned on with the advance of the nozzle, as long as an adequate vent hole can be opened. The members of the attack crew may be the ones best able to open a proper vent prior to operation of the fan. For those who don't advocate PPV, additional hoselines or negative pressure can be used. Large volumes of air can be moved with a fog nozzle. Situations like this may require break-apart nozzles or combination nozzles. The exit hole must be large enough to handle a volume of air from an entire hospital wing. Many of these buildings, especially prisons, have small windows, which may not encompass the area needed to pass a sufficient amount of smoke. An additional line may be required to protect the nozzleman, so fire won't be pulled toward him. Roof ventilation may be an option if conditions warrant.

Search will normally be limited to small areas once extinguishment and ventilation operations are already in progress. Occupants in the immediate vicinity of the fire may have to be

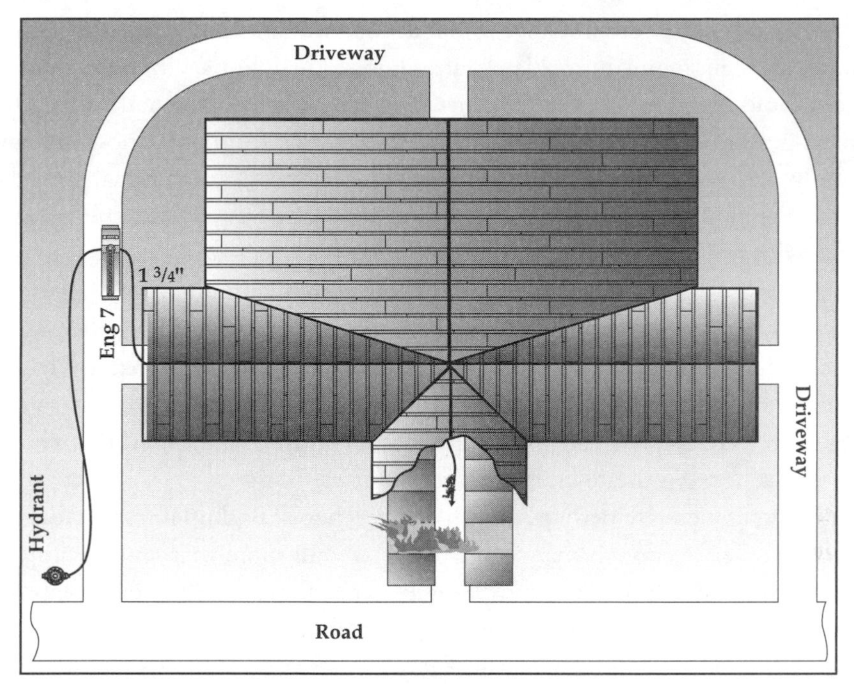

In a wing-type nursing home, the first line should be stretched to get the nozzle between the fire and the most victims.

Classroom layouts can differ from week to week in the same room. Movable desks will make search difficult. *Credit Coleman*

moved if they're alive. If they're not alive, other actions will take precedence. Search may also be required in adjacent rooms immediately after knockdown, but again, only those in imminent danger should be moved. After the flames have been knocked down, doctors, nurses, and other medical personnel may be permitted to assist in caring for those exposed to smoke and flames. Simply put, searches in a defend-in-place operation will be minimal, and the efforts at locating and accounting for the exposed will be more of a triage. After the triage, removal operations may begin.

Searches in schools can be difficult. Coordinated search with rapid ventilation is the best option. Ironically, any action that will reduce the time spent on search-and-feel will help to save life. If search is required in a classroom, for example, try not to move any chairs that are in place. Search between the rows, and don't throw or push chairs in front of you or to the side. Doing so will eventually create a bottleneck of furniture and make further search almost impossible. Move between the rows as easily as possible.

Evacuation operations are perhaps most likely in schools. Backup lines will facilitate those evacuations. If you have enough manpower to do so, pull more than one backup line. The first should shadow any unprotected crew. Search teams, for example, should be protected. If additional staffing isn't available, then a hoseline on each stairway will help facilitate the evacuations. Make sure that the lines don't impede anyone's progress. Throw the hose to the inside of the stairwell, and have a firefighter stationed where it crosses in front of the last step.

In penal institutions and hospitals, having a backup line will normally be a second thought. Depending on the amount of fire, these lines may never be pulled. Remember, I consider a backup line to be one that protects interior crews, not another attack line. The initial efforts in hospitals and prisons should be toward putting out the fire and ventilating. If search is in order, then the backup line should shadow the search team.

One of the toughest parts of being a chief officer at one of these fires is in finding the balance between aggressiveness in firefighting and concern for the safety of your crews. Trust me, a balance must be found. Crews will extend themselves to a greater extent in these buildings than they will in other types of occupancies. No doubt the life hazard has much to do with this. This is where establishing a rapid intervention team truly eases my mind as an incident commander. Having four members at the ready outside of a building allows me to extend those crews working inside just a little bit further.

SCENARIO

The school was built in the late 1930s and constructed of a variety of materials. The exterior walls have the appearance of true ordinary construction, but there's also structural steel placed vertically and horizontally and tied into the brick and block, adding mass to the structure. There are standpipes inside the building but no automatic sprinklers. The standpipe doesn't have a hose, just a connection.

Some classrooms have no desks. Kindergartens and lower-level grades may be set up similar to this. How would you search this classroom? *Credit Coleman*

An old parochial school sits behind a church. What construction concerns would be presented with this structure type? *Credit Coleman*

This older school will be the setting of the scenario for this chapter. *Credit Coleman*

The floors are supported by massive wooden floor joists laid on top of the exterior walls and also masonry, load-bearing interior walls. On top of the joists are wooden planks covered with poured concrete, and finally with hardwood flooring. The roof consists of wooden planks supported by horizontal rafters. On top of the planks is tar-and-gravel weatherproofing.

The school is for grades K through 8. It stands three stories above ground, and it has a full basement. The basement does have small slit windows.

There are twelve classrooms per floor. The first floor also has a group of offices occupying about a quarter of that level. Across the hall from the office, in the center of the building, is the auditorium, which has fixed seating. There is a full gymnasium in the basement, along with several storage areas and two industrial arts classrooms containing many power tools.

The second and third floors are laid out identically, with six classrooms at each end of a long hallway. A center stairway divides the floor, and there are also stairways on either end. Each room is approximately 24 X 36 feet, with an attached cloakroom of 24 X 6 feet. The classrooms have individual desks and no fixed seating for the students. There's also a teacher's desk in each room, although the teachers have the ability to position them and decorate their classrooms as they wish. For reasons of energy conservation, the windows were updated about thirty years ago. They now consist of insulated glass tile, with three small plate-glass windows set in. Hinged at the bottom, these small glass windows can be opened. The ventilation system has a pressure-relief system, which calls for large grates placed near the doors, allowing for the passage of warm air into the unheated halls. These grates will also allow smoke to migrate into the halls and thus throughout the building.

The fire occurs on the evening of the fall Open House program. After the parents leave, a disruptive and disgruntled child ignites some papers in a janitorial closet near the auditorium and on the first floor. This 12- X 30-foot room has shelves of floor sealant, waxes, and cleaning supplies. There's also a stockpile of paper products and rags. After he sets the fire, the miscreant runs back to his classroom and closes the door. The door of the closet, however, is left open, and flames quickly lap across the ceiling and out the door.

Small-Department Response

The incident commander should consider several factors immediately on arrival. The school is occupied, and the extent of the life hazard is an unknown. He has several firefighters on duty who have children in this school. In small towns like this, almost every responder knows someone associated with this school. The IC must therefore pay particular attention to the problem of freelancing and actions that stem from emotions rather than knowledge, experience, and training.

The school case study as seen from side "A" and "D".

If the same amount of fire were showing at 3:00 a.m., the IC would have automatically gone defensive, but given that there are people inside, he has no option but to go offensive. The question is whether to go little or big, but the correct choice is to go big. Even with reduced staffing, his objective is to knock down the fire or hold it where it is. The lines that he places won't be mobile. They'll be brought in from two directions, and they'll be stationary once charged. He commits the entire first-alarm contingent to the attack, and he assigns two paramedics on the ALS unit to begin a search. Ventilation will be left to units on the second alarm. If the weather is warm, this building may be difficult to vent, since it's common for schools to have openable windows. The truck crews will spend much of their time closing windows in the unaffected portions of the building so as to be able to control the draft created by their fans. Positive pressure will likely be the quick vent tool of choice. If the fans can be directed toward the auditorium and can help hold the heat and fire to that wing, then they'll have done their job. As venting improves the conditions within, members from both the first and second alarm will join in a thorough search.

If the fire is such that crews must operate on multiple floors or over large areas, the IC must consider deploying a backup line. Two members are enough to handle it if manpower is truly at a minimum. The IC must also consider establishing a RIT. Depending on the circumstances, this may at times be a primary assignment, and at other times it may be a luxury reserved for the second alarm.

The school in the case study as seen from Sides "C" and "B".

The auditorium must be overhauled. Crews of the second alarm will initially help with this. As time goes on, these members should be the first to be released. Many auditoriums, of course, have high ceilings, and great care must be taken when overhauling the upper portions. Depending on the construction, some of this may be done from above, and hydraulic overhaul is a possibility. This probably isn't the time, however, to practice tying off an erect ladder at four corners—the so-called dome raise that we elders learned as recruits.

Because of its suspicious origins, this fire will also require extensive investigation. Many chief officers aren't trained in collecting evidence, documenting a crime scene, and interviewing suspects. This is a time to get professionals involved. As an incident commander, I have never had the need to assign a fire investigation sector. If I were to assign this task, I would place it under the Planning section, since documentation and other related aspects of an incident are already under this section.

Medium-size Department

In a medium-sized department, the first-in officer is less likely to be faced with the problem indigenous to small towns; that of having their own children in the school. This also goes for the members of his crews. Still, he'll be confronted with another problem. The bigger the town, the greater the chances of having a sizable mob of civilians demanding that firefighters get in and do something.

As in the earlier scenario, the crews of the first alarm should deploy big lines. This tactic cannot be overemphasized. If in doubt, go big. The incident commander will need help in determining how many people are missing. The police can be very helpful in this regard, and using them won't take away from ground forces. Horizontal ventilation is again accomplished with PPV once the attack lines are in use. A midsize department is more likely to have thermal imagers, and search operations can be greatly enhanced by using this equipment. Backup lines must be considered if crews are operating on multiple floors or if the fire takes hold of stairways or halls. These fires also call for the establishment of a RIT as soon as a crew is available or procedure dictates.

Overhaul and determination of the cause will top the list of secondary assignments. Additional crews can be called in for the overhaul. The initial crews at this fire mounted an aggressive interior attack, and they're likely drained. Let the fans blow for a while, and be sure that the school is well vented prior to commencing overhaul. Two firefighters on the inside can watch for flare-ups in the meanwhile. This is the time to regroup and for the IC to have an informal staff meeting, at least with his officers. The meeting might encompass a tailboard critique and a quick debriefing of the initial crews. If the cause of the fire is known, I believe that those who waged the battle deserve to know, too. If the information is sensitive in nature, tell them so.

Because more manpower is available, more operations can be accomplished concurrently. The IC will also be able to tend to certain command activities, such as hospital notification and establishing an information section. The incident commander will also want to brief school officials on their options. For example, they'll need to know how and when the investigation will take place. They can be kept informed about overhaul and salvage. If the school will have to be closed for a time, the IC can also involve the media. Remember always that firefighters come in contact with fires hundreds of times a year. The average citizen may only personally be involved with one or so in his entire life. The stuff that firefighters take for granted is very new to the public. No matter how routine their concerns may seem, they should be kept informed and encouraged to ask questions.

Large-Department Response

Given the uncertainty as to the number of victims and the extent of the fire, the initial commander should request additional help immediately on arrival. Facing no shortage of resources, the incident commander may mount an attack and begin to search at the same time. Expecting that the outside vent man and the forcible entry team of the first due truck will complete their preassigned tasks, the IC can assign other truck members to search. The first and second engines, meanwhile, pull large lines to the fire, and the third engine takes a line above the fire. Rescue personnel will also go to the fire floor or the floor above to search for fire and life.

Second-alarm crews will be used to relieve members whose bottles have gone off, as well as to complete secondary assignments. Rapid intervention teams will be assigned from the second-alarm truck crew. Once the fire has been knocked down, tower ladders can be used to overhaul from above, or smoothbore nozzles can hydraulically overhaul from below.

Chapter Eleven Questions

1. The key to fires in institutional occupancies is that the residents inside are under somewhat ____________ circumstances.
2. The key problem with fires is institutional occupancies is one of __________.
3. To control this concern, a ______________ strategy is usually warranted in fires in institutional occupancies.
4. The normal victim/rescuer ratio should be considered ______________.
5. What is the key to a defend-in-place strategy?
6. _______________ will make or break a defend-in-place strategy.
7. Who usually is in the best place to open the proper vent hole for PPV?
8. When can staff and doctors be used to assist in victim removal at a hospital or nursing home fire?
9. With defend-in-place operations, search will be more of a __________ than actually removing victims.
10. As it relates to school fires, what could be the best option as it relates to search?

Questions for discussion

a) Discuss the legal and moral issues of "defend-in-place" operations.
b) Discuss the victim/rescuer ratio in institutional occupancies.
c) The author believes that the key to successful search in a school fire may be aggressive ventilation efforts. Discuss this statement.

Chapter Twelve

Fighting Fires in Lumberyards

To my father, the word *lumberyard* meant only one thing: a place of business consisting primarily of an array of open-air sheds, built of wood and containing piles of wood, plus stores of various other building materials. To modern firefighters, the word still retains this connotation, but it can also refer to various warehouse-style retail stores. The home-improvement chains literally offer one-stop shopping for the both the home handyman and the professional tradesman, and they present fire problems both similar to and different from those of traditional lumberyards. This chapter will attempt to consider both types.

The potential for a conflagration is present in an older lumberyard. These fires can generate tremendous levels of heat. The ability to extinguish a fire in this fire usually depends on the ability to throw enough water to stifle the heat. Fire spread can be rapid, and even light winds can pose an enormous problem. The flame front can propagate itself by both convection and radiation, and burning pieces of wood and char can be lifted by convection currents and float for blocks until the current loses heat and allows the brand to fall back to earth or onto a rooftop.

The route of access to an older lumberyard can range from difficult to impossible. In most of these facilities, the lumber is stored in the rear of the property, enclosed by fencing and often bordered by railroad tracks. The space between the rows of lumber typically isn't wide enough to accommodate fire apparatus. Sometimes only a forklift can make it down these aisles. Venturing inward with aerials and other apparatus may prove difficult and, once in the yard and committed, secondary movements of apparatus may be nearly impossible. If you're going to commit vehicles to the compound, position for a single placement. Spot the apparatus in the best location and then dump water.

Normally, life safety isn't a problem at a lumberyard fire. You must account for the employees, but most of these businesses operate with a limited number of people.

An older Lumberyard. Drive-through buildings and many sheds are present in these occupancies. *Credit Coleman*

Fires in building-supply superstores have many of the same characteristics, but they're also more problematic in that they're confined. These stores carry everything from lumber and roofing materials to paint and PVC, and the heat produced by a good-sized fire can be overwhelming. That's the bad news. The good news is that, because of their size, these buildings are normally required to have sprinkler systems. Still, life safety is more of a concern in these environs, since the customers are free to roam throughout the store. In a traditional lumberyard, customers typically make their purchases at a front desk, then drive around to a designated spot inside the gate to receive their goods from yard workers.

The main, front-office building of a lumberyard is usually of ordinary construction or wood frame, though some of the newer versions have steel walls. In the north, these walls may be insulated, as with other occupancies in the area. Walls of ordinary construction may or may not be reinforced. Concrete masonry units are probably the material of choice. Most of the brick that you find will only be veneer over CMV, lending a dressier look to the exterior. If the main office is a wood-frame structure, it's probably built that way to showcase some of the merchandise inside. Expect false fronts, knee-high walls, and windows that are

This Home Depot represents the more modern lumberyard springing up over the country. More than 100,000 sq. ft. of combustibles and flammables under one roof. *Credit Coleman*

backed by plywood. The roof assemblies will normally be of wood or wooden truss. The rafters will normally be of 2 X 8s or larger, depending on the span, or older wooden trusses. Bowstring trusses aren't uncommon in some lumberyards in some areas. Plywood or purlins over the rafters, with wood shake or asphalt shingles, will top off the roof assembly. With the exception of the trusses, these roofs should be similar to those on homes in the area.

The home-supply superstores have exterior load-bearing walls of steel or CMV. The steel types are held in place by steel columns set in concrete. Walls of concrete masonry units will normally be reinforced due to the size of the structure. Again, some may have brick veneer fronts, but the side and back walls will be of naked CMUs. During your preplanning, you may also come across tilt-up concrete walls on these buildings. The roof assemblies are predominantly of steel-bar joist, with Q decking as sheathing. Weatherproofing is provided by tar and gravel or sometimes a built-up roof. Due to the spans involved, girder-and-column assemblies of unprotected steel are common. Anticipate partial collapse in well-involved fires in these occupancies.

Because the sales area must be able to carry heavy loads, expect to find concrete-slab floors in both the building-supply superstores and the main offices of most older lum-

Lumber piles under the protection of shed roofs are present in traditional lumberyards. These piles present extreme exposure problems at fires in these occupancies. *Credit Coleman*

beryards. These buildings weren't made to have underground storage. Although some do exist, basements are rare.

The principal concerns with the main offices of older lumberyards center on the fire loads inside and the voids created by the facades. Fire will spread rapidly throughout these occupancies. The voids will trap the heat and fire, disallowing the movement of air that usually exists in the presence of dormers and windows. The outbuildings are usually open-air structures and highly susceptible to fire. Collapse will be a certainty in well-advanced fires and, in a worst-case scenario, a sizable fire in a lumberyard will result in a conflagration.

In full-scale fires in the warehouse-style lumberyards, collapse due to excessive fire load and lightweight construction will be the main concern. Ventilating the roof will be difficult and dangerous due to the steel-bar joists. Well-advanced fires can be produced under the roof assembly in metal-deck or built-up roof assemblies. By code, floor areas of more than 12,000 square feet must be sprinklered, and a building-supply store of 100 X 120 feet would be relatively small by today's standards. Expect sprinklers, but supplement them early, since you should also expect the sprinklers to be overwhelmed.

Firefighter injuries can be significant at a lumberyard. Operating in narrow aisles amidst piles of lumber banded together by metal straps, coping all the while with numerous trip hazards below, instantly puts members at risk. Exhaustion enters quickly into the equation, since these fires generally require long hoselays. Get relief units in early, and set up rehab no more than thirty minutes into the incident.

Overhaul will also be a problem. You may need to overhaul large piles of burning lumber hydraulically. Aerial platforms and master streams will prove very effective for this. You may need to bring in bulldozers to turn over large piles of debris.

Whether deploying for a traditional lumberyard or an indoor superstore, keep in mind the old axiom to confine, control, and extinguish. The first three to five minutes of the firefight will set the tone for the rest of the incident. The initial hoselines must be placed so as to confine the fire to the smallest area possible. If you arrive and the fire is in the rear of the yard and the wind is pushing it toward the front, then stretch the lines from the front with the thought of meeting the flames head-on and beating them back toward material that has already been burned. Conversely, if the fire is working from the front to the rear, go around to the back to stop the advance. Set secondary lines from the flanks to keep the flame front from moving backward.

A fire involving storage sheds at a lumberyard can generate great amounts of heat and significant exposure and brand problems.

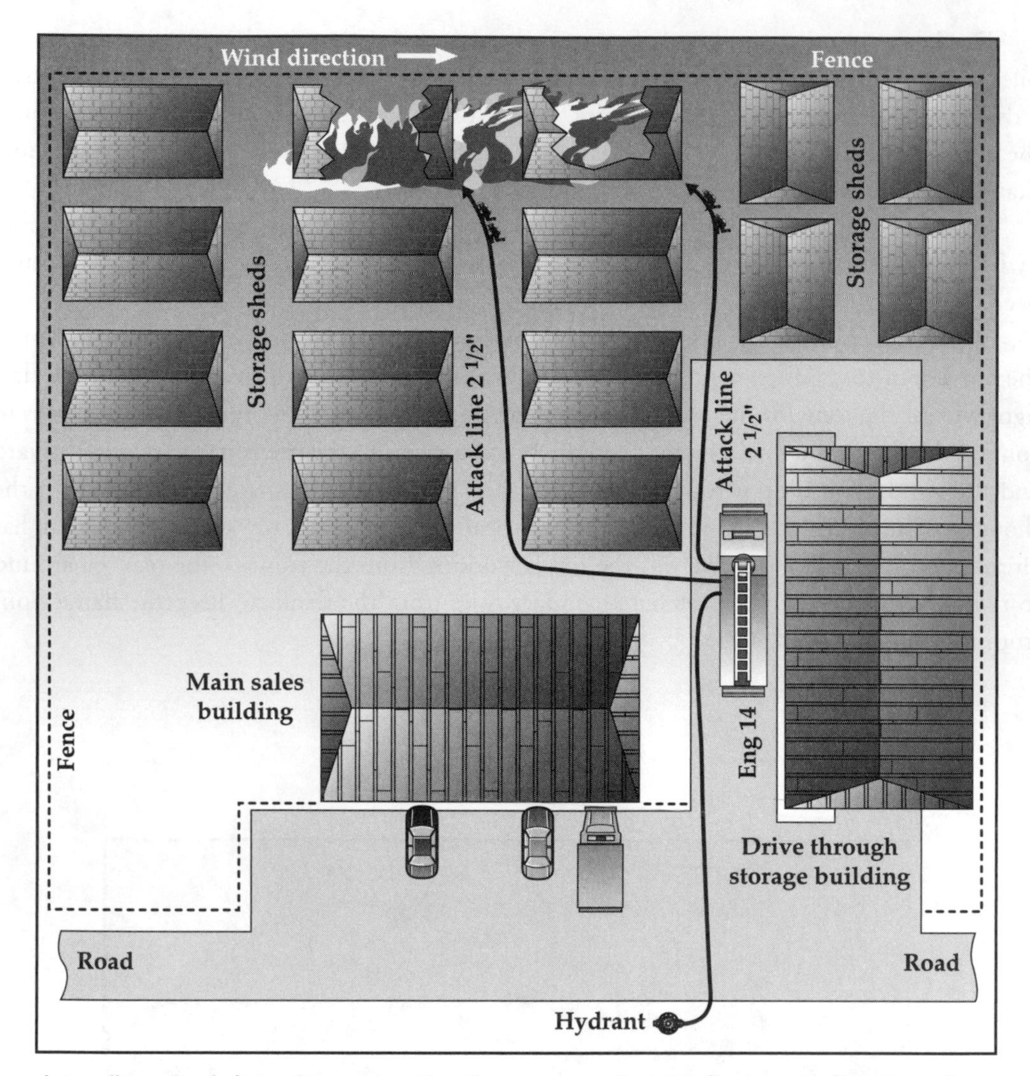

A two-line attack from the proper direction ensures that the fire is cut off and confined to the smallest area possible.

Manpower is critical. My rule of thumb, which can be substantiated by no scientific fact or process, is to triple the resources that you initially think might be needed to handle the incident. Suppose you believe that the flow of two aerial streams and four handlines can hold the fire in front of you. Two engines will be needed to supply the two aerials. Flanking the fire will require one engine on each side of the fire, each flowing two 2 1/2-inch handlines. That's four engines and two trucks already committed. For me, that's a regular alarm plus one truck. In my world, I would ask for a third alarm immediately. This gives me a second-alarm assignment to assist at the original fire, plus a third-alarm contingent to deal with spot fires and exposures. Stage the leftover units, and assign them as necessary. Given a decent wind (there always seem to be strong winds at lumberyard fires), an advanced fire in an outdoor lumberyard can move several feet a minute, and who knows how far those flying brands and embers might go? Get additional crews on the scene early, and hopefully you can send them home clean.

The officer assigned to the initial attack should be thinking about two things: the direction of the attack and the amount of water needed. If the fire is moving from west to east, the first lines should come in from the east, but this isn't always possible. Fences, barbed wire, railroad tracks, and other obstructions may block the way. Sometimes wind, radiant heat, and flying brands will ward off a frontal attack. In such a case, two simultaneous events must occur. One is the positioning of aerial streams in the forward position to hold the fire in check. Second, large-caliber handlines should flank the fire, also to stop the forward spread.

Another example of a modern lumberyard. Expect long hallways with significant fire loads in these buildings. Sprinklers should hold in check incipient fires. *Credit Coleman*

Lines of 2 1/2 inches equipped with smooth-bore nozzles will give the best balance of reach, flow, and maneuverability while pushing the fire the least. Train this line as close to the front as possible and work backward to the point of origin. Forget about smaller, more maneuverable lines. The hose that you pull should be large and able to flow a minimum of 250 to 300 gpm. Maneuverability isn't as important as initial placement and gallonage. If you can't stop or hold the progression of the flames within the first few minutes, you'll probably be in for

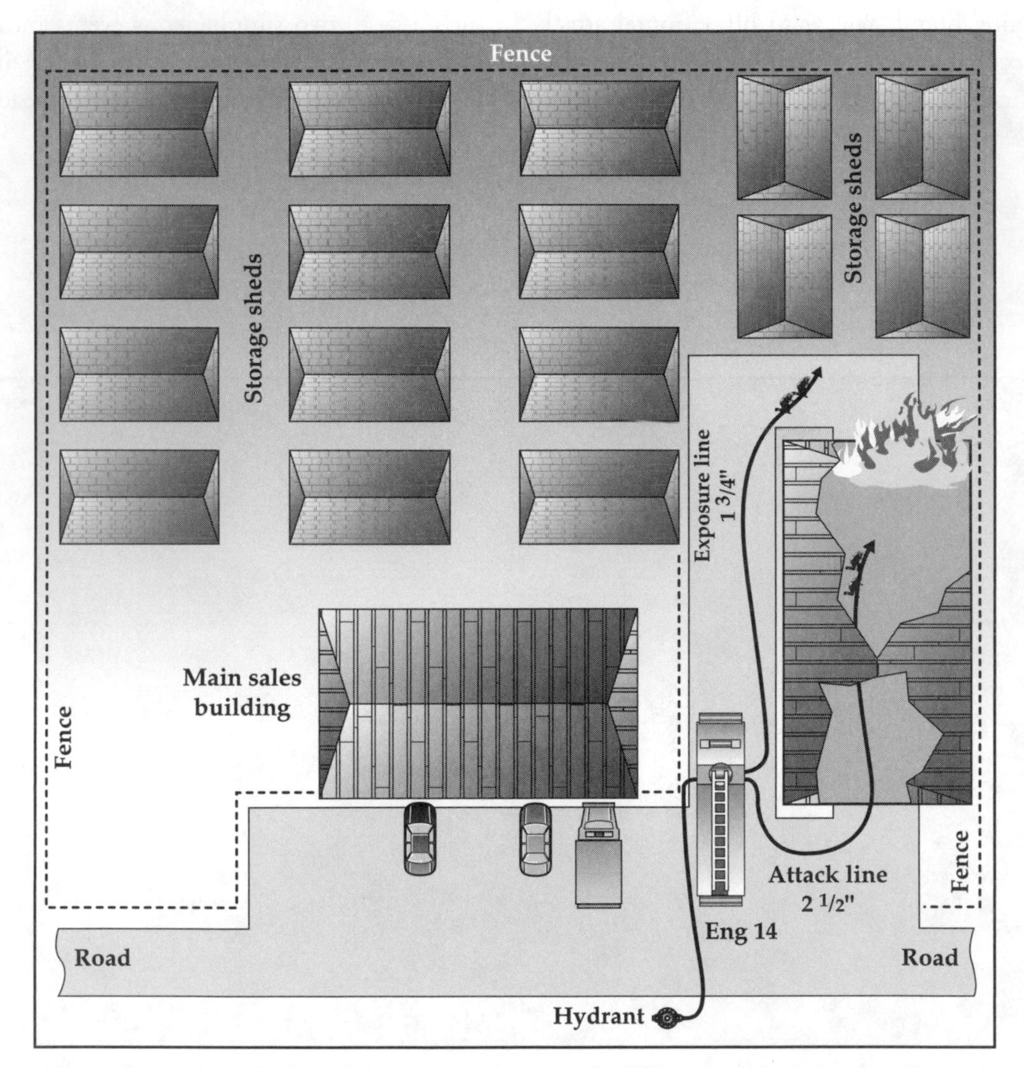

When a fire involves interior buildings at a lumberyard, 2 1/2" lines provide the greatest "maneuverable" fire power. This line is again, taken in to confine the fire to the smallest area possible.

the long haul. In a lumberyard, you only get one chance to stop the fire before it gets out of control. Pull a large line and knock the heck out of it.

As indicated above, search will normally be a secondary assignment at an older lumberyard. The mode will be similar to that of a defend-in-place strategy at an institutional fire. The initial efforts are all focused on attack. Still, the incident commander should attempt to account for the employees and customers as soon as possible. Older lumberyards don't bring in hoards of customers as do the indoor superstores. Limit the first searches to the main building. Input from the owner should be useful in determining the severity of the search problem. The IC should have the owner or manager perform a head count of his employees as soon as possible. If an employee is missing, a single crew can begin to search the yard. As ever, the search should start as close to the fire as one might expect savable victims to be, then to work toward unaffected areas.

In a home-supply warehouse, search is another matter. With dozens of aisles and hundreds of cars in the parking lots, significant search operations are required. Hopefully, automatic sprinklers will restrict the area where flash fire is a possibility, but it isn't hard to overwhelm the sprinkler system in this sort of occupancy. At minimum, a life or search rope is required, but a hoseline is better. These huge buildings can be full of turns and dead ends. Take a dry line and charge it when you reach the end of your search area. Work the aisles off the sides of the hoseline. The aisles don't move, and neither should the hose. In these occupancies, use teams of two, even with the oriented-man type of search.

A fire in the main office of an older lumberyard will probably be best vented by whatever means your department uses to open up single- and multifamily residences. Be on the lookout for decorative windows, dormers, skylights, and so on, since no appreciable air movement will be gained by venting impermanently mounted display items! Healthy flames in heavy fire loads can create strong convection currents, translating into a significant accumulation of heat in the upper levels of the structure. If topside ventilation can be done safely, go for it. Otherwise, wait for the roof to vent itself.

Ventilation will be tougher in the retail warehouse. Heavy fire loads will bring temperatures at the ceiling to high levels very quickly. Such heat may create a self-sustaining fire in the metal-deck roof, and this will be very difficult to extinguish. This sort of fire can only be fought from the underside. Any water applied to the top will simply wash off without effect. Large-caliber streams directed at the ceiling from below will be required to cool the burning gases.

Steel-bar joists fail in five minutes at approximately 800°F. That's not a very hot temperature to expect of a fire in one of these occupancies. Although there usually won't be a general failure of the roof system, you should anticipate sectional collapses. Let the roof open itself and extinguish through the vent hole from the underside with master streams. Set monitor nozzles inside doorways and other safe areas remote from the main fire, then back out and wait for them to have an effect. Smoke flows from areas of high pressure to areas of low pressure; i.e., from where the fire is, out the vent hole. By allowing the roof to vent itself, you create a natural vertical channel for smoke and heat to follow, thereby stemming the horizontal flow out of doorways and windows.

A look inside a modern lumberyard. Note the high rack storage in these occupancies. *Credit Coleman*

More than likely, backup lines will be an afterthought at a working fire in the outdoor portion of an older lumberyard, though one will be required if the fire is in the main building. A backup line may be warranted in the yard if the crews are advancing between rows of lumber that could become involved. The size and position of the line will be similar to the deployment regime used for a vacant commercial occupancy or industrial emergency. Don't place it too close to the working crews—just close enough to guarantee their quick retreat, if necessary.

Don't even think about fighting a fire in an indoor superstore without a backup line. Given the size of these stores, the backup officer may want to increase the size of his crew to position a firefighter every fifty feet along the line. This will help maintain communications, and it will maximize the view of the avenue in and out of the building. Reports and updates must be forwarded to the backup officer as the fire progresses.

Some may not see the need to establish a RIT at an exterior fire, but one is needed just the same. With tremendous water flows and raging fires, bundles and stacks of lumber can all too easily become unstable. It doesn't take much to trap or pin a firefighter in cumbersome gear. There have been incidents at which firefighters, clouded in by dense smoke, have gotten lost in the mazes created by lumber piles. Get a RIT established at any working fire in a commercial occupancy, indoor or outdoor. These members don't have to come from the first unit on the scene; rather, this is a good assignment for mutual-aid forces.

At a superstore, the RIT officer may have to be concerned about tens of thousands of feet worth of floor space. Think about it. How could you effectively cover such an area? Suppose a crew becomes disoriented inside. You didn't even know that they were in there, and the officer of the apparatus has no idea as to their whereabouts. What would you do? Where would you start?

Three answers come to mind for this issue. First, don't let your department get so lax at fires that this can occur. Second, if it does occur, talk to the crew on one radio frequency, and have them describe what they see. Have them stand motionless and let their PASS devices activate. Three, split your RIT crew into four teams, with one trained RIT member as the leader of each team. Let these teams work inward from four different directions until they can hear and locate the PASS devices. I can't think of much else that can be done. Don't pull out any crews working the fire. Let them keep the flames in check and hopefully snuff out the source of the problem.

SCENARIO

The fire occurs in an old lumberyard on a Monday evening around closing time. There's a main building that contains the offices, a sales counter, and retail space for tools, hardware, and other items. Outside, in the back, there are two large drive-through outbuildings. There are also ten or twelve storage buildings, open on one side, housing lumber, truss assemblies, and the like. The entire compound is fenced in.

The incident starts as a vehicle fire inside one of the drive-through buildings. A yard worker grabs a fire extinguisher, but the flames and smoke drive him back. The worker runs into the main office to call the fire department. The owner of the vehicle remains in the yard.

Small-Department Response

The first officer on the scene from any department must weigh the capabilities of the initial response against the scene confronting him. In this case, there's one outbuilding in a major lumberyard with heavy smoke showing. The officer believes that his crew and the next-in engine can quickly get in and knock the devil out of it before it gets beyond their means.

With the fire confined in one building, this may be an instance in which it's more advantageous to pull smaller lines, given the shortage of manpower. If two 1 3/4-inch lines can be placed in service faster, their mobility within the structure might outweigh the firepower of the 2 1/2-inchers, and certainly reach won't be a consideration. For safety, the IC might also order the deployment of a PPV fan, to be placed behind the attack crew. With the fan run-

A look at one of the drive through buildings in the scenario. *Credit Coleman*

The fire for this case study as seen from side "A" and "D" as crews approach.

ning, the likelihood of fire dropping down behind the advancing crew will be lessened. The fan shouldn't even be turned on, however, except on word from Command or Attack.

Depending on the arrival of other units and the size of the fire, of course, the IC should be able to maintain an offensive attack. If no other units arrive in due time, and if the flames still aren't under control, he should probably pull the crews out and regroup to protect the

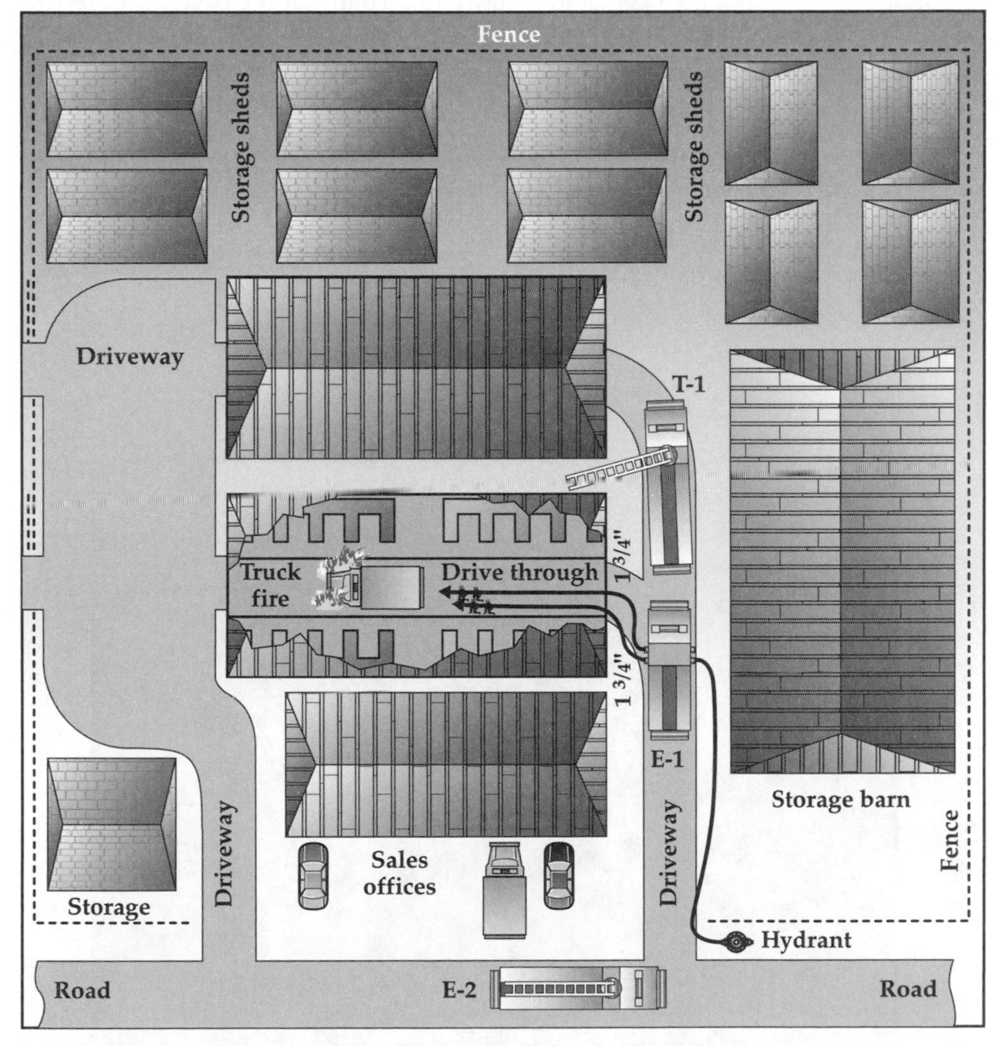

Initial line placement of Engine 1, Truck 1 and Engine 2. Engine 3 will lay-in and set up a monitor nozzle between the two drive-through buildings.

exposures. In most cases, with no life hazard, the contents of a storage building in a lumberyard just aren't worth the potential cost to firefighters.

If the crews are eventually driven out by heat, the IC could set up aerial devices at both ends of the storage building. He could also establish handlines and deck guns. Because of the potential, he should order a precautionary third alarm. There's no disgrace in being driven back if an offensive strategy fails—not if the initial attempt was a reasoned one and the crews were pulled out in a timely manner.

The first item on the list of secondary assignments should be search, given that there may be savable victims on the fireground. If this isn't likely, then secondary searches can be conducted in conjunction with overhaul. Head counts among the workers and customers will help. Count the vehicles in the parking lot and try to account for the owners. If every owner can be identified, chances are better that there are no victims among the ruins. The police can help in this task.

Heavy equipment will help greatly with overhaul efforts. Steel buildings often sag down on top of their smoldering contents. Such rubble piles need to be torn apart and cooled off. Many lumberyards already have heavy equipment on-site. The condition of the initial crews

A close-up of the drive-through building involved at this case study.

is always a concern. Make sure that they're up to the task of overhaul. If they're not, get relief crews to the scene and send the first crews home.

Medium-size Department

With more manpower available, the first-in officer might opt for large-diameter hose from the outset. Command has the second-in engine assist with the stretching of the line. This provides enough firefighters to maneuver the line, plus an officer in charge of the crew. In total, he lays two attack lines and a backup line for the first phase of operations. If the ini-

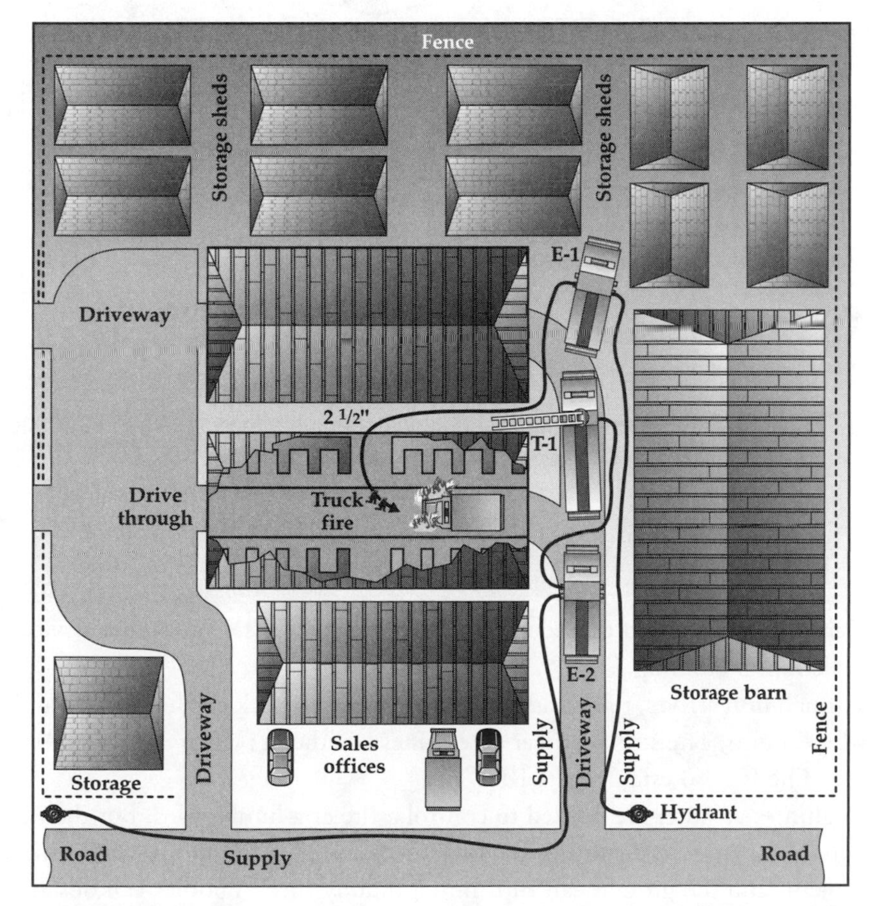

In the Medium-size department scenario, the first line is a 2 1/2" line taken in to confine the fire.

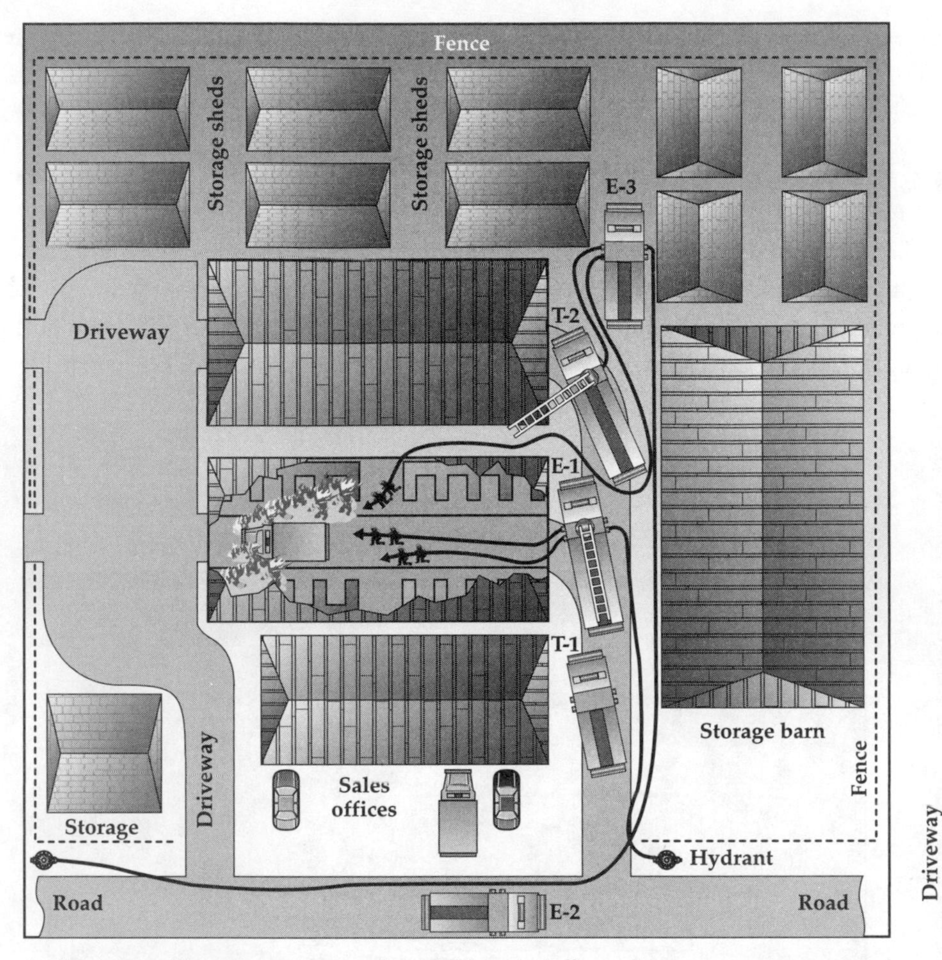

In the Large-Department scenario, additional staffing allows crews to surround the fire with the first alarm response.

tial attack crew can't hold and then darken the fire, then he'll set up master devices to hold the fire to the initial building.

As the squad unit arrives, the IC assigns its members to check on the workers and account for their whereabouts. The search officer determines whether a primary or a secondary search is warranted. The IC also establishes a RIT.

Great volumes of water are needed to control a fire at a lumberyard, but the more water that you throw, the more difficult overhaul will become. Flooding in the yard can become a serious problem, and the ground can turn into a muddy mess. Puddles can obscure numerous trip hazards, and foam can hide still more. Beware of piled material that may become unstable. Any baling straps must be cut with care, if they're to be cut at all.

Large-Department Response

The water supply and staffing won't be a major concern for the first-arriving crews, but you should sound a precautionary second alarm anyway. The crews will aggressively attempt to fight this fire.

The first and second engines both get a line into operation, and the crew of the first-in truck commences a search, based on information they gather at the main office. The third engine sets up with the second truck on Side D of the fire building in case the interior crews can't hold and darken the flames. Additional lines are brought from Side C as backup lines. Generally, the crews of the second alarm are set up to protect exposures, and one engine is used for brand patrol. A RIT is assigned as procedure dictates.

Chapter Twelve Questions

1. The potential for _____________ is present in an older lumberyard.
2. What two means of heat transfer are factors at a lumberyard fire?
3. Access is generally not a concern to the rear areas of an older lumberyard. True or False
4. The decorative fronts of the main sales-room of old traditional lumberyards are not only decorative but also always functional. True or False
5. What are the two key factors to firefighters' injury at fires in old traditional lumberyards?
6. The first hose lines at a lumberyard fire should be placed so as to _____________.
7. The author gives a simple rule for staffing at a lumberyard fire. It is _________ the normal staffing requirements.
8. If the attack can't be mounted from the front of the fire, then _________________.
9. The officer assigned to attack should be thinking about what two things?
10. ________ lines with _______ give the best balance of reach, flow and maneuverability.
11. When searching a new home supply, the author suggests you use the _________ as a point of reference.
12. In a fire in a built-up or metal deck roof in a new retail lumber warehouse, large caliber streams directed _______________ will be required to cool the burning gasses at the ceiling.
13. Steel bar joists begin to lose strength at ___________.
14. Back up lines are not needed for fires in the yard or an older lumberyard. True or False
15. RIT is not normally required in any exterior fire. This holds true in yard fires in older lumberyards. True or False

Questions for discussion

a) Discuss the factor of exhaustion to crews at lumberyard fires as it relates to "the rules of engagement".
b) How would your local police department work as a tool for determining the location of workers at a lumberyard fire? Would they be a help or more of a hindrance and would they know ahead of time exactly what you wanted of them?
c) What concerns would the RIT officer have in a fire in the yard of an older lumberyard?

Chapter Thirteen

Fighting Fires in Enclosed Malls

An enclosed mall is essentially Main Street without traffic. It's a multilayered string of retail stores and restaurants under a single roof, with limited means of access. Typically, malls are laid out with several wings that lead either to an entrance or an anchor store. They tend to be of newer, lightweight construction, although an trend has developed over the past decade or so of remodeling old factories, warehouses, schools, and other fallow buildings into shopping malls.

The pedestrian walkways of newer lightweight enclosed malls are usually complete with fountains, gardens, pushcart merchants, and rest areas. The average retail store can range from less than 200 square feet to more than 4,000 square feet. The anchor stores customarily rise to at least two levels and may be 50,000 to 100,000 square feet per floor. The smaller stores are typically defended after hours by roll-down security gates. The anchor stores tend to have traditional glass doors and foyers at their outside entrances and roll-down gates only at their interior entrances.

Converted enclosed malls differ in that they have fewer anchor stores, perhaps none. The myriad retail shops are normally divided by interior partition walls, or they can be free-form marketplaces and have few dividing walls at all.

Given their size and local codes, most enclosed malls are protected by automatic sprinkler systems, and some have standpipes throughout. The standpipe systems can be vertical or horizontal. Still, given the ceiling height in some of these buildings, you may encounter an advanced fire due to delayed activation of the system, as well as fires in areas remote or hidden from the sprinkler flow.

The degree of life hazard will vary with the time of day, as well as the season. Obviously, late-night fires pose less of a threat than those that occur during business hours. Fewer employees and shoppers will be present both early in the day and near closing time. Some

The main entrance to an Enclosed Mall. *Credit Coleman*

stores do 60 percent or more of their annual sales volume during the December holiday season. Generally, the life hazard will be highest around the winter holidays, in the spring, and when kids are returning to school. Volume decreases during the off-season months of June, July, and October. The fire loads in these occupancies also varies greatly, depending on the merchandise at hand.

In the older, converted malls, the exterior load-bearing walls are usually of brick, block, or mixed construction, often in combination with steel. In buildings of ordinary construction, expect the interior columns to be either of ordinary construction, steel, or a combination of the two. Some storefront taxpayers with masonry load-bearing exterior walls, as well as party walls spanned by simple wood joists, have been converted into malls, and many times the party walls have been breached to enlarge the shopping floor.

In newer, lightweight malls, expect the exterior load-bearing walls to be of steel construction, concrete, or a combination of the two. These buildings are fire-resistive and, by code, should be protected by automatic sprinklers if they're larger than 12,000 square feet. Larger buildings may be unsprinklered, however, if there are rated fire separations that reduce individual areas to less than that size.

This old Civic Auditorium once housed traveling circuses and plays. Today it is occupied as an Enclosed Mall with antique stores, a factory glass outlet and many restaurants and small shops. *Credit Coleman*

If the exterior load-bearing walls are wholly of concrete and the building is only one story tall, suspect tilt-up concrete construction. These buildings are essentially held together by steel cable, which can fail at around 800°F. Steel exterior load-bearing walls may be covered with siding of steel, brick, aluminum, concrete slabs, or even wood.

The roof of a converted mall can be one of several types, depending on the age of the building. Older structures can have tar and gravel over plywood or planks. You may find newer rubber or membrane roofs over planks or plywood. Be careful of added dead loads, such as air-conditioning units. Try to determine whether the joists were bolstered to handle the additional weight.

Newer lightweight malls have metal-deck roofs, or perhaps membrane or rubber. These will normally be over steel-bar joists. Any air-conditioning units on newer malls were probably in the original plan and, in all likelihood, will already be supported appropriately.

The floors in older conversions are usually of wood or concrete, covered with linoleum, tile, carpet, or painted concrete. Newer lightweight malls have floors of concrete slabs or Q decking supported by steel-bar joists. The final covering is usually of carpeting or tile.

Conversions can present construction concerns for firefighters. Rarely do architects factor in the various stresses associated with fire emergencies. Many of these buildings aren't even

A converted Enclosed Mall. Would you expect this building built in the 1940's to be sprinklered? *Credit Coleman*

In many older towns across the country, Old store-fronts such as these have been converted into Antique Malls. Passageways have been added from building to building creating maze-like conditions. *Credit Coleman*

designed for the loads and stresses placed on them as a result of renovation. Facades and various interior modifications may create an aesthetically pleasing environment, but they may also place new burdens on old and sometimes saggy supports. Electrical and plumbing systems brought up to code may abrogate the compartmentalization of individual units. Unprotected openings can also plague these buildings. The ease of access from store to store translates into easy routes for fire and smoke. Newer malls also suffer from this problem. After hours, the various stores may only be protected by security bars or acrylic, and the steel supports of the building may be exposed to the flames.

Serious fires have occurred in enclosed shopping malls. Well-involved fires in these structures require extensive resources to control. An abundance of water will be needed to absorb the heat, and there can be no shortage of manpower. A large number of firefighters may be required to search an anchor store of 100,000 square feet and fifty percent involvement on a floor loaded with clothing racks. Heavy smoke conditions mandate that we feel for victims rather than simply look for them, and we have thirty-minute SCBA, which will last for about fifteen minutes under these conditions. Even a small fire may require over thirty firefighters just to man the hoselines, augment the sprinkler, vent, perform search, effect rescues, and pull backup lines. This figure doesn't include the incident commander, safety officer, or the RIT. Those thirty would be just about able to handle a small incident, comprised of three handlines (twelve firefighters), four members to vent, six to search (two teams of three), four to perform rescue, and six to man the backup lines. A fire that involves half a floor of an anchor store could require fire flows of more than 16,000 gpm to control.

For the incident commander, the primary strategic consideration will probably be life safety, and the second will be extinguishment. In most situations where there are known victims inside, and assuming that the sprinkler system is incapable of handling the flames, the IC will have to consider an aggressive fire attack. Nozzle placement will have to be correct if the responders are to confine the fire successfully and provide the maximum edge for civilians.

Following the attack, ventilation becomes the priority. Two distinct concerns present themselves at this juncture. The first and foremost is life safety. The second is the need to minimize fire loss, which can be considerable even to occupancies not directly involved in the fire. Controlling evacuees is another problem, for malls necessarily entail large numbers of both employees and passing civilians. Controlling the rescued and self-evacuated means having a place to put them, not to mention the help of the police and a little planning. To let them wander about, leave, or even go back into the structure is only to ask for trouble. Neglecting your obligations with regard to civilians may subject you to scrutiny and criticism by the media, not to mention the city fathers. Round up and control your evacuees. Use buses, unoccupied buildings—whatever is available, but keep close tabs on them.

It's true that most fires in these buildings are small and rarely beyond the capacity of sprinklers and a simple mop-up line. Still, fires of larger proportion do occur. There are two considerations when confronted by a working fire in an enclosed mall. The first has to do with the strength and volume of the line. A 2 1/2-inch line with a smooth-bore nozzle may be required. I say "may be required" because chances are that this line is overkill, but you should have such capability with you just in case the fire is larger than you thought. The other consideration is the avenue of attack. This is one instance in which the positioning of the line has a clear bearing on the number of persons protected and the amount of property saved. Make every effort to get a line between the victims and the fire. The moth-to-the-flame syndrome, i.e., stretching directly to the fire, can often translate into needless loss and injury.

For fires located in anchor stores, take the initial line through an adjacent door. Most anchor stores have several exterior doors. If the fire appears to be seated on one side, go around it and come in from another side. This simple action should be enough to help you halt the spread and shield the victims.

It's to your advantage when the fire is in one of the smaller stores along the mall itself. Taking the line in from the central walkway should get protection between the victims and the fire. Many of these fires are started by arsonists in fitting rooms. Others start in the back of the store, in the stockroom. Fires in the back room are generally accidental, resulting from careless smoking or misuse of electrical equipment.

This stand-alone department store should be considered similar to an Enclosed Mall as it relates to strategy and tactics. *Credit Coleman*

An Enclosed Mall with high ceilings, exposed truss assemblies, and kiosks lining the wings. *Credit Coleman*

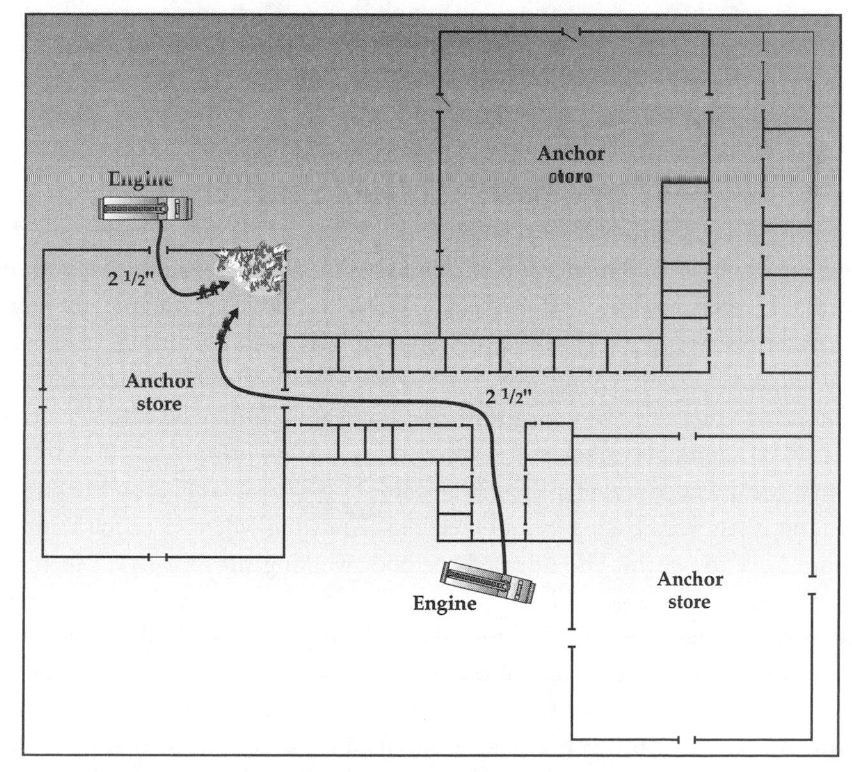

For fires in anchor stores in Enclosed Malls, initial lines should be taken in to confine the fire to the smallest area possible.

When the fire occurs along the pedestrian walkway, take in the lines via the closest means of access. Employing an indirect attack with smooth-bore or straight streams directed at the ceiling will inhibit spread, and it won't introduce the element of push that results from attacking the fire directly.

Many fires are reported to be out or held in check prior to arrival. Other times, firefighters are told that the sprinklers are flowing water. Don't be lulled into a false sense of security be either of these reports. Pull at least a 1 3/4-inch line to control any fire that the sprinkler head can't hit, such as under clothing racks. Fires in apparel and other merchandise can generate a lot of heat. It looks bad if we walk in with nothing, only to have to run back out again for a line.

There is some good news about fighting fires in these structures. By code, new enclosed malls are built with HVAC systems that keep the smoke from banking down to the floor. Ventilators and ejectors in the HVAC system are designed to keep the smoke several feet from the floor, in fact, even in working fires. Additionally, backup generators are required in the event of a power disruption. Thus, ventilation should not initially be a major concern at a fire in an enclosed mall. To be sure, we'll need to supplement the HVAC system at a major incident. There may be smoke pocketed in areas over which the HVAC system has little control. Still, the system may fail. It is imperative that you determine ahead of time how the system can help you at a fire, as well as plan how you'll vent in a worst-case scenario. Don't think that a design advantage will mean the end of your ventilation worries at a fire in an enclosed mall. Always consider what will happen if the system shuts down. Venting an enclosed mall can be no big deal. It may also prove to be the most restrictive strategic factor at a fire. Be prepared for both.

Most of these sprawling structures have atriums, and some have skylights. Some have strips of glass all along the ridge of the roof. Take the roof and take the glass. This isn't as easy as it sounds, and the relatives of civilians killed by falling glass will find high-powered lawyers to file negligence suits if you go about it the wrong way. Get someone below you, preferably with a thermal imager, and ensure that the fall area is clear before you take out any glass.

The ventilation system may make search as simple as walking through the mall looking for victims, but searching an anchor store can require literally dozens of firefighters if the area is full of smoke. If you can only accomplish a partial search, it may be better to commit your on-scene crews to aggressive attack and leave search for other units.

As always, commence a search as close as possible to the fire in an area where savable victims might be found, then work back to remote sections. Knowing where to search leads you to the next decision: how to search. The oriented method, working off of a 2 1/2-inch line, is my method of choice. In such large environs, using walls and other fixed points for orientation will be difficult at best. A hoseline stretched into the occupancy, however, will provide both a convenient reference and a traceable route of escape. Stretch the line dry and then charge it at the order of the officer of the search crew. The oriented man will sit on the line and maintain contact with his crew as they canvass the areas to specified distances on either side, using left- and right-handed searches, as appropriate. After the crews have worked back to the entrance or an area of clear visibility, the line should be drained and moved to another location.

To discuss search is to discuss two constraints: staffing and time. Staffing must be considered if the searches are to be done expeditiously enough to save anyone. In a large area, it would be reasonable to expect that two firefighters and an officer can search no more than about 5,000 square feet in fifteen minutes. This is an open floor area with few or no obstructions. Given a plethora of display racks and wrap stands, the area that you can cover will be substantially reduced. In an anchor store, you'll need more than twenty crews to cover 100,000 square feet effectively. Time being the second constraint, I'll give civilians fifteen minutes of search time in a primary search mode when the smoke is heavy. After that, secondary searches will probably be warranted. In lighter smoke conditions, the interval can be extended. Survivability increases when the visibility is better, and self-rescues become more likely.

I often speak to departments all over the United States. One comment I often hear from volunteers is that there isn't enough to train on. They complain that they need new "stuff" for drill night. Try this! How would you search a mall in your jurisdiction? How would you search the school gymnasium involved in a flash fire during the middle of a basketball game? Think about it, talk about it, then try it. Large-area searches will drain resources faster than any other fireground operation.

Parking lots along with illegal parking can present access problems at Enclosed Malls. *Credit Coleman*

Backup lines are a necessity for mall fires, and they play a dual role. The mission of the backup team is to protect the interior crews and ensure their route of egress. In a mall, the backup may also be needed to maintain the evacuation route for civilians. Thus, the backup officer must establish coverage between the civilians and the fire as well as between the members and the fire. This is accomplished through proper line placement, as well as discipline. "Proper line placement" for a backup crew means directly behind the interior crews. To be effective, the backup team cannot move ahead of the suppression team. Backup isn't recon. These members aren't looking for fire, they're looking for victims. They should be behind the working crews. The best route for the backup team to follow is along the attack line. If the attack team takes its line through the main door of the mall, then backup should follow accordingly. Deploying this way lessens the chance of winding up with opposing lines. It also guarantees that the location of the attack crew is known, and it ensures that the crews won't be outflanked by the fire.

If a search crew is inside, then these members become the first responsibility of the backup team. If more than one search crew is working, either on another floor or within an expan-

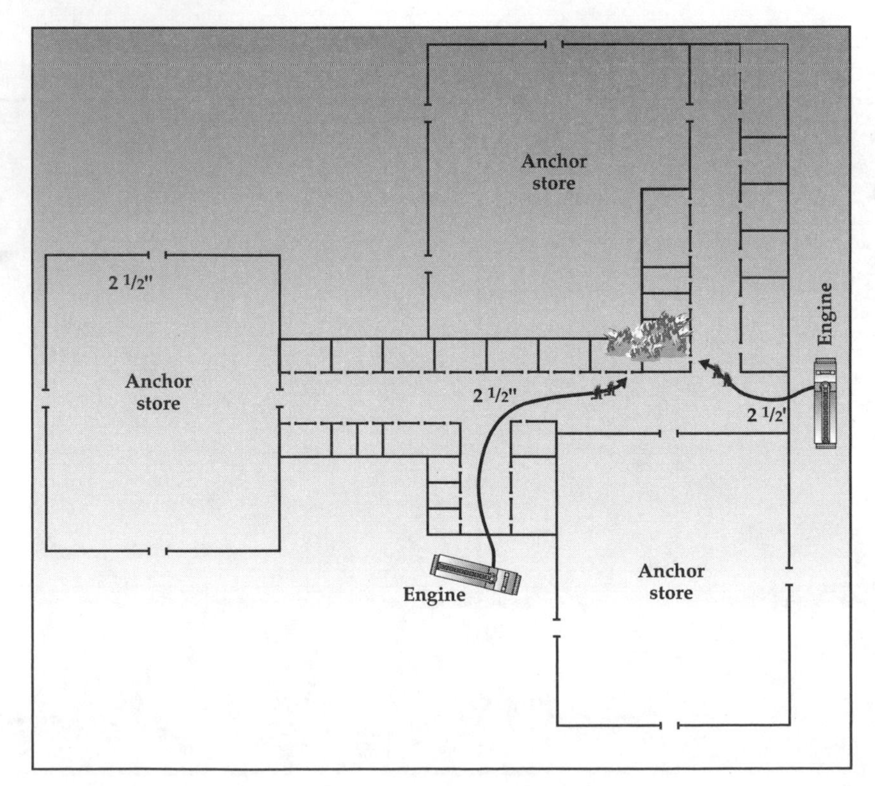

Positioning of a back up line in an anchor store fire.

sive area, there should probably be more than one backup crew to shadow them. The size and location of the fire will have a great bearing on the number of search and backup crews needed. If the officer in charge of backup feels that fire conditions warrant a separate backup line for the search and attack crews, then the IC should be notified to make the final decision.

The officer and crew of the backup team need discipline so that their focus won't become blurred. The greatest temptation for a crew with a large hose is to put out fire, especially when a large flame front is right in front of them. The officer needs to ask himself, "If I were on the attack line, where would I want the backup team to be, and what would I want those members to focus on?"

Given the size of these buildings, the RIT may need to stage at an area remote from the command post. If the fire is in a store somewhere along the mall itself, these members may wish to stage in an anchor store entrance, for example, out of the smoke. Staging at such a location may make it impossible for these members to monitor the accountability board. The usual 360-degree walk-around of the fire structure will be required, with the lion's share of attention going to construction concerns. Ingress/egress isn't normally a concern at emergencies involving enclosed malls. Converted enclosed malls probably pose more problems for a RIT than the newer versions, and the vast array of construction types is principally at fault for this.

Expect significant search problems in these occupancies if the ventilation systems fail or can't keep up with the products of combustion. *Credit Coleman*

The members of the RIT should use one-hour SCBA bottles for extra work time. If the RIT officer doesn't feel comfortable deploying into a dangerous area inside an enclosed mall, he should ask the IC whether any other crews with hoselines are in the vicinity of the firefighter in distress. If the answer is no, the officer might request a backup crew with a line to cover his own team. If the IC denies this request, then the RIT officer has one terrible decision to make.

In such large buildings, a support team may be needed to shuttle in bottles to the RIT. If the RIT members are using sixty-minutes bottles, then the support team should start inward to meet them no more than twelve minutes after the bottles have been activated. On each trip, the support members should take in one bottle for each member of the RIT, as well as bottles for themselves. After the exchange, the support crew should head back to the entrance from which they came. In the event that they're delayed, they'll each have another bottle to use on their way out. If the travel time for the RIT from the safe zone to their target area is more than fifteen minutes, there'll be problems. The RIT members will exhaust their air in about thirty minutes. The support crews shuttling air in to them will need at least thirty minutes' worth of air. That's fifteen minutes in and fifteen minutes out, and that's cutting it close! To eliminate this problem, each member of the RIT support needs to carry two or more bottles. Harnesses and slings have been developed that allow a firefighter to carry four SCBA bottles at once. Prepare for these eventualities before the Mayday call goes out. Have your contingency plans at the ready.

An EMS branch will be required when confronted with an advanced fire during business hours at an enclosed mall. The officer in charge of this branch will work for the operations officer. The EMS group will be responsible for establishing triage and treatment for injured civilians, as well as transportation to medical facilities. If staffing allows, one paramedic should be assigned to triage and another to treatment. A firefighter with a portable can run the transport part of the operation. There should also be a staging area for the ambulances, and each victim must be tracked so there is no confusion as to who winds up where.

Like other aspects of the incident, salvage is a major issue at a fire in an enclosed mall. Significant loss of property can be averted if merchandise is protected as soon as possible from the effects of fire. An abundance of salvage covers, fans, and other items will be needed. Even so, salvage operations should only commence if there are sufficient crews available and the other necessary aspects of the incident have been addressed.

SCENARIO

The city has just acquired its second enclosed mall. It's located on the north side of town, in an area annexed from a township two years ago. The city is expanding at a prodigious rate—no doubt faster than the size and capability of the fire department that must protect it.

The mall has four anchor stores. Each is two stories tall, with approximately 100,000 square feet of space per floor. Each has center escalators, as well as an elevator located near an exterior exit. One of them has a full restaurant.

Fighting Fires in Enclosed Malls

This Enclosed Mall will present several distinct problems in the scenario. *Credit Coleman*

In design, the mall itself has three single-story wings, joined at a food court at the hub. Each wing is lined with small stores and has kiosks and other features along the way. The anchor stores sit at the end of each wing.

The exterior load-bearing walls of the anchor stores are of tilt-up concrete. There is also an enclosed skeleton of steel to help support the second floor and roof. The upper floors are of corrugated steel decking with a layer of poured concrete. The steel-bar joists of both the upper floors and roof are unprotected throughout the mall. A rubber membrane comprises the uppermost layer of the roof. The entire complex is built on a slab. The interior walls are of steel-stud construction, and there are unprotected steel columns within the wings.

The mall has a full sprinkler system, and there is a pump house located on the exterior. The HVAC system is designed to keep any smoke ten feet off the floor. This sophisticated system, complete with two backup generators, requires a twenty-four-hour engineer to monitor it. The generators are located in the sprinkler pump house, and they also serve the fire pumps for the sprinklers. The mall's backup generator is in a similar building on the opposite side of the lot.

The fire starts on a Saturday evening at 2055 hours. The summertime temperature is 87 degrees with 90 percent humidity. The air-conditioners of the mall have been running at capacity for more than three days. There is the threat of a thunderstorm. Lightning has been flashing across the sky for twenty minutes, but as yet there's been no rain. The mall is due to close in five minutes, and the parking lot is emptying. Several high school kids have met at the food court and are deciding what to do with the rest of the evening. The gates of the various stores are beginning to roll down to the halfway position, and the cashiers are beginning to close out their registers.

A semi carrying a load of merchandise for one of the anchor stores swings into the parking lot. The driver has been hung up in traffic for the past few hours. He knows that he's late. If he can get to the store before the guys on the loading dock leave for the night, then he can still head home with an empty trailer. Otherwise, he'll have to spend the night in the sleeper cab. He's driving faster than he should be. As he rounds a turn, a car with two older women fails to stop at a stop sign. The truck driver swerves to avoid them, jumps the curb, and stuffs the tractor into the pump house. The car carrying the two women is pinned under the trailer. Someone calls 911 about the accident, and the dispatcher upgrades the response to include ALS because of the report of people trapped.

Small-Department Response

There's a reason fire codes require heavy fire protection for these types of occupancies. Simply put, they can generate devastating fires. We must know and understand the applicable codes and also have contingency plans ready in the event that a worst-case scenario presents itself, as it does in this scenario.

From the outset, the incident commander must have an understanding that something is not right. Fire suppression systems should be controlling almost any fire in this type of occupancy. Once he comprehends that the system has failed, the incident commander must decide whether to launch an offensive or defensive strategy. Obviously, given a small department, staffing will be a concern, and the IC must call for additional alarms early if those units are to arrive soon enough to be of any help.

Without an interior fire suppression system in place, his crews will be struggling to fend off a fast-moving fire. A small-department IC would likely be justified in giving up the entire store so as to save the rest of the mall. He should set up a strong defense in advance of the fire and wait for the flames to come to him. Using a two-front defense, he could establish interior monitor nozzles just inside the wing attached to the involved anchor store. These will be used to prevent any drop-down fire from spreading into the one-story wing. Outside, the IC sets up four trucks, two on each corner of the wing, to suppress any fire and attempt to hold the flames to the anchor store. Two other aerials will attempt to darken down the fire where it has vented. These two aerials set up near the wing and assist in holding the line.

Water is a prime concern. With the master streams in place, the flow will be up around 7,500 gpm. That's the capacity of the five engines, plus another engine supplying the sprinkler system. If this sort of gallonage isn't available, the IC will have to prioritize his streams.

The case study for this chapter involves a fire at an Enclosed Mall. This fire was started on Division 2 of this JC Penney store by a lightning strike.

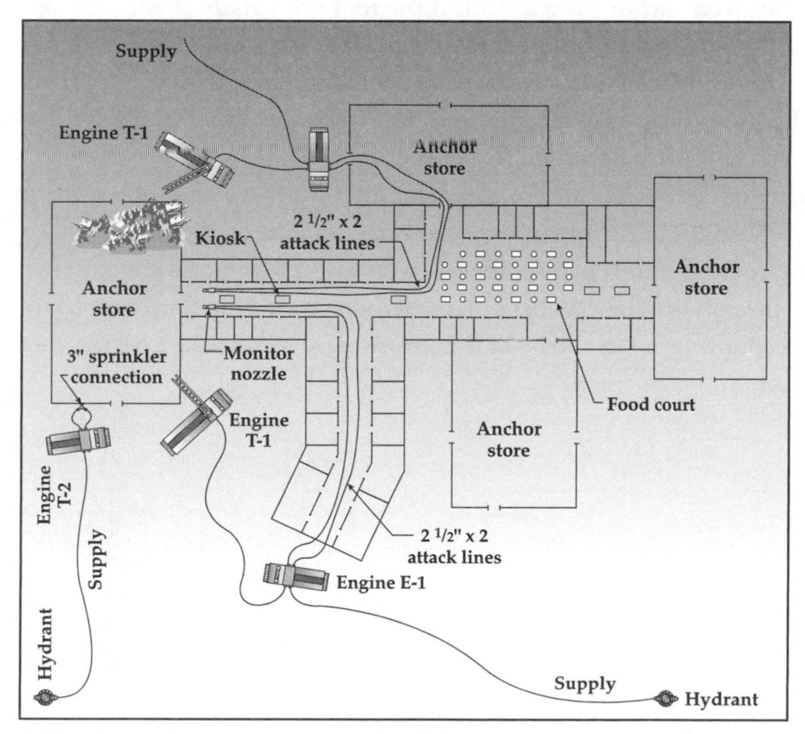

Initial line placement in a Small-Department response. As crews arrive, more monitor nozzles and aerials will be set up.

The interior crews will have to give frequent updates as to their progress. If lines aren't needed inside, that will help the exterior effort. Again, priorities should be based on saving as much of the mall as possible.

Assuming that sufficient water is available and that the streams darken down the fire, an accounting of all store personnel, civilians, security guards, and cars in the parking lot will be required. Police and security personnel can help with this task. The police may have to run the license plates to identify the owners.

Overhaul will be a major issue. In an occupancy such as this, a private salvage company may be needed to go through the ruins. Once the fire is out and all of the hot spots have been extinguished, and after the investigation has been completed, the building can be turned over to the owner. It will be up to the owner to hire a salvage company to determine what can be saved.

Medium-size and Large-Department Response

Given a sizable contingent of personnel, the incident commander may opt for an aggressive interior attack. A monitor supplied by large-diameter hose would be the tool of choice. Big water should be available all around the mall. A typical layout would be three monitors and four 2 1/2-inch lines in operation on the second floor, with the intention of cutting the floor of the involved anchor store in half. If the roof vents itself, it will alleviate a lot of heat on the second floor. Units from a third and fourth alarm can establish defensive, holding streams outside of the wing leading to the anchor store. In the event that the interior crews can't hold the fire inside, these exterior lines will come into play. As always, having a RIT is essential when crews are committed to the interior.

As the fire darkens, the interior crews should be able to shut down the monitors and advance their handlines on hot spots in the far reaches of the second floor. The initial crews can be rotated into rehab, as members from the third and fourth alarms take their place. Even though the fire has been darkened down, it's no time to relax on the inside. Naturally, accounting for civilians and performing overhaul will be the most important secondary tasks to be assigned.

Chapter Thirteen Questions

1. Anchor stores in enclosed malls can range from _________ to more than ________ sq. ft. per floor.
2. Very few enclosed malls are protected by automatic sprinklers. True or False
3. What type of construction is used in most converted malls?
4. The two biggest factors in estimating the potential for life hazard in an enclosed mall are __________ and ___________.
5. Lightweight construction can be expected in new enclosed malls. True or False
6. For the IC, the primary strategic consideration at a fire in an enclosed mall will be ________.
7. The moth to flame syndrome can have negative effects in fires in enclosed malls. True or False
8. HVAC systems generally have little positive effect on working fires in enclosed malls. True or False
9. The atrium glass may be effective as a means of ventilation in an enclosed mall. True or False
10. What are the two constraints as it relates to search in an enclosed mall?
11. Generally, one back up crew can cover up to three search teams in a working fire in an enclosed mall. True or False
12. The ____________ will have the greatest bearing on the number of search and backup crews needed.
13. RIT will always stage at the Command Post at a fire in an enclosed mall. True or False
14. If available, a ________ SCBA bottle should be used at a fire in an enclosed mall.
15. If many civilians are injured at a fire in an enclosed mall, an EMS under _________ may need to be established.

Questions for discussion

a) Discuss the constraints of staffing as it relates to search in an anchor store.
b) Discuss initial line placement in an anchor store as opposed to a fire in one of the stores in a wing in the mall in your district.
c) Would your department have the staffing to not only fight the fire but also salvage the merchandise in a large mall with heavy smoke and water damage? If not, what are your options?

Chapter Fourteen

Fighting Fires in Restaurants

The term *restaurant* in this text will refer to any commercial establishment that prepares food, whether its customers are able to consume their meal on the premises or not.

Life safety and rapid fire spread are the two main problems in restaurant fires, and rescue operations may prove difficult. Kitchen floor plans and congested seating areas make for close-quartered search. Tables, chairs, carts, appliances, and a host of other items all create an environment of movable obstacles that will hamper search. Personnel operating in storage areas and kitchens will have to contend with everything from limited working space to mountains of combustible stock to slippery floors. The interior decor, the presence of flammable liquids, and the numerous utility connections associated with cooking can all conspire to create a nightmare scenario beyond ordinary expectations. Recognize also that many fires in restaurants are deliberately set by owners intent on collecting insurance money. Such fires may be given a head start with accelerants and be unusually advanced by the time initial units arrive. Naturally, the life safety concerns in a restaurant will vary with the day, time, weather, and business hours of the establishment.

In a large city, where most citizens walk or use mass transit, you may have no cars in a parking lot to count, and consequently no idea as to how many victims might be inside. Some restaurants are protected by modern sprinkler systems, and some may not even be equipped with handheld extinguishers. Panic can be a factor among the victims. Usually the patrons only know one way out—the way they came in. Terrified, they will converge on a single exit, creating a bottleneck that too often leads to injury and death. Firefighters sent in for search and rescue may be put at risk. They may be denied entrance; they may be mobbed and left disoriented; they may be trampled. Simply put, the fire may only be the catalyst that fuels a more pressing problem.

There are many Restaurant chains across the country that pose significant fire and life-safety problems. *Credit Coleman*

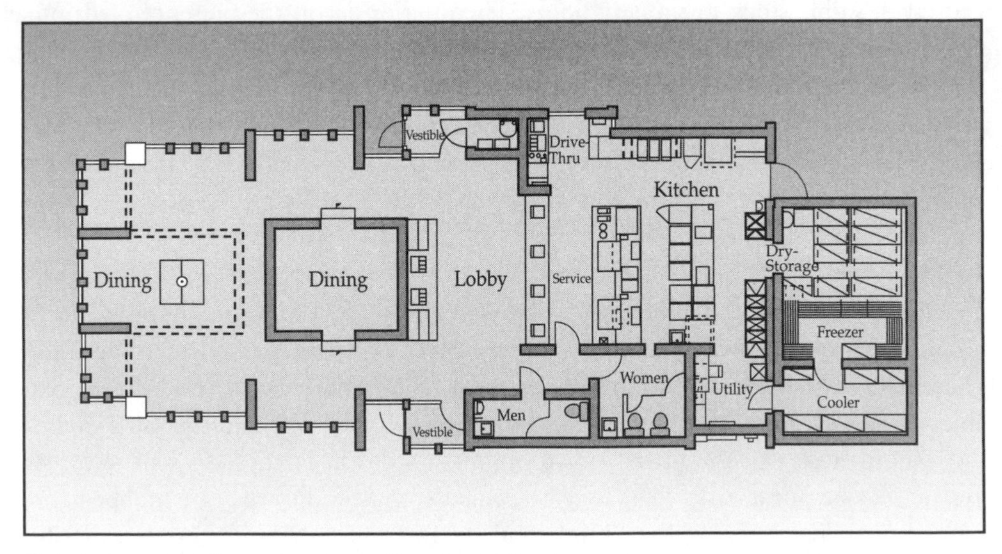

This is a typical floor plan for a "fast-food" Restaurant.

Many Restaurants are housed in buildings that were not originally designed as eating establishments. Construction concerns may play a significant role in fires in these occupancies. *Credit Coleman*

In terms of construction, restaurants are as varied as their fare. They may be housed in anything from new lightweight construction to converted heavy-timber mills. The smallest in town might be a 10 X 20 CMU hamburger stand; the largest might be an elephantine banquet hall; the highest might be sitting atop a steel-skeleton high-rise—and it might be rotating! Pre-incident planning is the only real way to find out what you'll need to know at the next emergency. Like other types of buildings, the roofs on older establishments are likely to be of wooden rafters, wooden planks, tar, and gravel, or perhaps rubber membrane or poured concrete. Among newer establishments, you'll find an abundance of lightweight trusses holding up metal decks, solid metal, asphalt shingles, membranes, or built-up roofs. The floors will normally consist of wood joists, wood trusses, or concrete.

Some of the most serious loss-of-life fires have occurred in restaurants. The Beverly Hills Supper Club fire in Kentucky comes to mind. The exact potential for loss of life will be affected by several factors: the size of the establishment, the time of the fire, the construction features, the presence of automatic sprinklers, the size of the department, and the level of training of the firefighters.

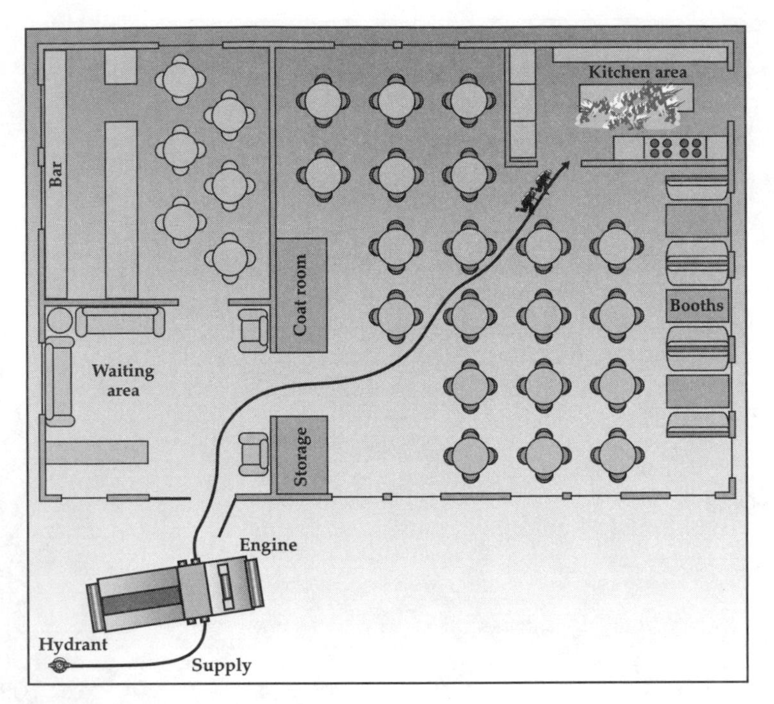

Initial attack lines should be taken in to cut off the fire's spread.

From the outside, would you expect apartments or more dining rooms on the upper portions of this old Restaurant? *Credit Coleman*

In determining his strategy, an incident commander's main concern is best expressed as a question: "Can we simultaneously put out the fire, vent, and search for patrons, or should we defend in place?" There are six principal elements that will help determine his answer. These include the character of the fire, the size of the restaurant, the number of patrons, the ability of the patrons to rescue themselves, the size of the responding forces, and the level of training of the department. If the fire is small, under control by automatic sprinklers, or in an area remote from the dining area, suppression and rescue operations can be launched concurrently. If the fire is sizable and the victim-to-rescuer ratio is at all lopsided (say, three to one or greater), then defend-in-place operations may be in order. The size of the response is certainly a key issue. Multiple alarms may need to be struck immediately. In rural areas, this means mutual aid and longer response times. There should be no delay in marshaling an army of appropriate size.

When launching an attack, go big and go early. If, on arrival, you're torn between search, ventilation, and other operations, tell yourself to put the fire out. This is especially important if staffing is a concern.

Suppression operations in a restaurant should follow the common attack principles. Place the hoselines so as to confine the fire to the smallest area possible. Put another way, lay the hoseline between savable victims and the fire. This accomplishes two objectives. First, it actu-

This old Restaurant will pose a significant search problem under fire conditions. Getting inside before the fire will reduce search times significantly. Know your buildings! *Credit Coleman*

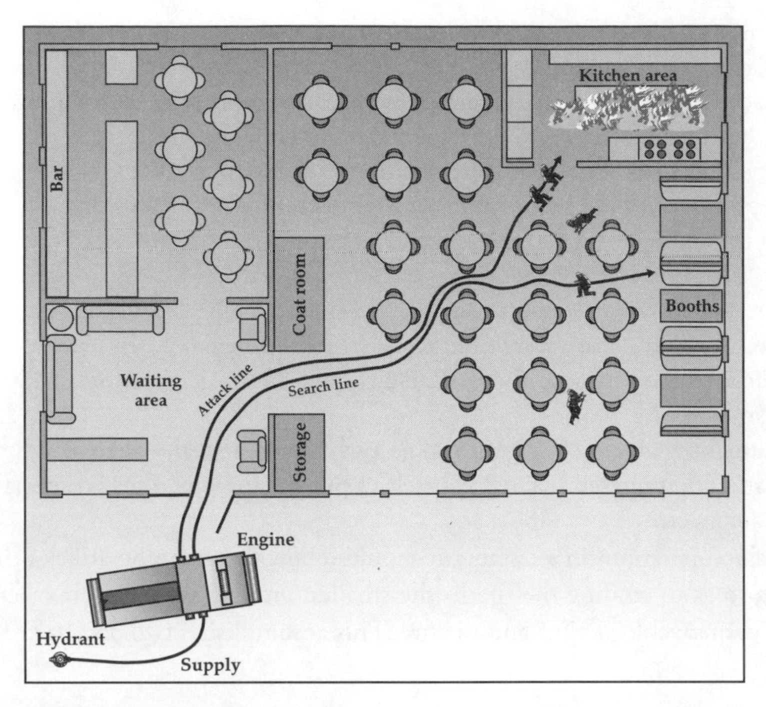

When searching in larger eating establishments, Search may want to search off of a 2 1/2" line.

Many larger cities have small Restaurants scattered throughout. You could find anywhere from none to 30 or more victims in an occupancy of this size. *Credit Coleman*

ally does confine the fire. Second, it gives civilians the best chance of survival. If you take in the line with the intention of attacking quickly and with no thought given to cutting off the spread, then you'll risk pushing the flames toward civilians, and you'll also soon be playing catch-up against an extending front.

If you foresee using flows of more than 200 gpm, then pull a 2 1/2-inch hose as your initial line. Take it in dry if staffing is low, then charge it and hit the fire from a safe distance. It'll take more members to stretch two 1 3/4-inch lines (four minimum) than it will to stretch one 2 1/2-inch line (two minimum) and, under most circumstances, you'll be able to flow more water with the larger item. An involved room of 20 X 30 feet will require a flow of 200 gpm, and that's not a very large room for a restaurant. An area of 40 X 40 feet will require more than 500 gpm, which can easily be accommodated by two 2 1/2-inch lines. Unless properly pumped, three 1 3/4-inchers are needed to flow that much.

What constitutes an effective search in a restaurant, given that there may be large numbers of civilians inside? Suppose the parking lot of a fast-food restaurant contains thirty vehicles. If each car is considered to have brought an average of 1.5 people to the premises, and given that a few may have walked there, we can expect that about fifty people were inside at the time the fire broke out. The question is, how many firefighters will it take to locate and remove even 25 percent of that number within fifteen minutes?

There should be no argument that, if we are to be successful at this incident, a substantial portion of the responding force must be committed to attacking the fire. In a typical response, you might have two engines, a truck, and a heavy-rescue squad, or about sixteen firefighters, plus a battalion chief and a driver. The attack team will require one engine company, or four members. Ventilation is also a must, and probably PPV, so figure one member to open the hole and one to operate the fan. I would also order a backup line inside to protect the interior crews, an assignment that will require two members. There are now eight members left for a search-and-rescue operation. Could eight members effectively search for, locate, and remove twelve or thirteen individuals from a burning restaurant, and can we consider one-quarter of the potential occupancy to be a good success rate? Should all eight members deploy inside to search, or should four remain at the door, acting as a rescue group?

Although stated rhetorically, the implications should be fairly clear. If you can't throw at least eight members into a search effort, then maybe you should divert all of your efforts into attacking and venting the fire, then mounting search operations as soon as the smoke begins to lift.

The overriding constraint on a searcher, of course, is the size of the air cylinder on his back. In Toledo, we use thirty-minute SCBA bottles. We have determined that the average firefighter will last fifteen minutes on such a supply, though warning bells have gone off in as little as ten minutes after deployment. Although studies on this and other tactical evolutions are still underway, I feel very comfortable in stating that an average firefighter can search a 1,000- to 1,500-square-foot area within that fifteen-minute span, given a floor area devoid of unusual obstructions and surprises.

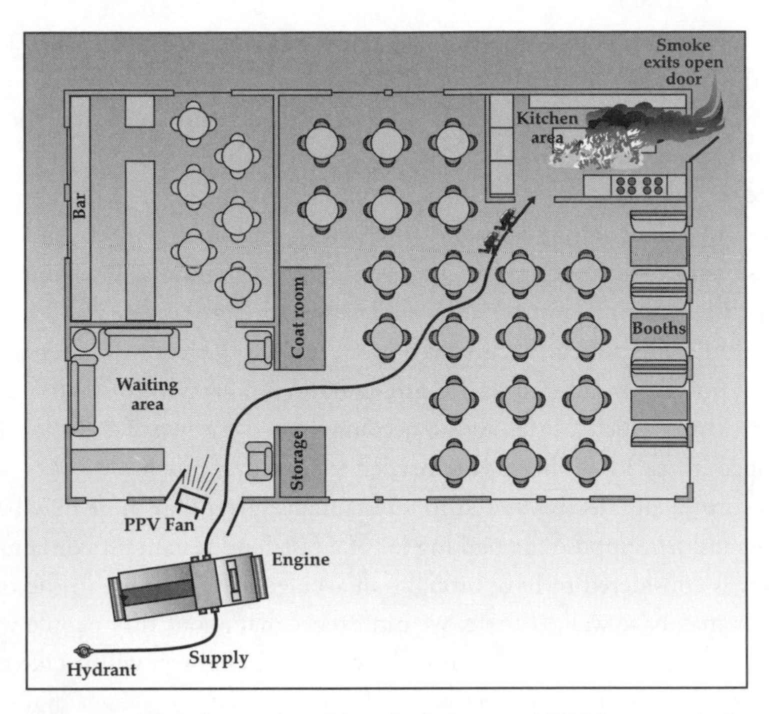

After the fire is located and knocked down, a PPV fan can aid in rapid ventilation efforts.

Search should take less than two minutes and under most circumstances be handled by the Attack crew at this small national chain Restaurant. *Credit Coleman*

It would be nice to say that each car in the parking lot indicates 1.5 patrons inside the building, but there really is no reliable formula for determining just how many people might be inside, and sometimes it's downright impossible even to guess. The seating capacity in one of the modern chain restaurants can range from fifty or so to several hundred and (except in a fast-food or greasy-spoon environment) people seldom eat alone. In heavy smoke conditions, we'll probably have no more than fifteen minutes to locate viable victims. If we take longer than that, we'll probably be calling the coroner. A restaurant that covers less than a thousand square feet is a lot to canvas in a short time, especially if there are victims to be found and, in the aftermath of panic, tables and chairs are strewn all over the place.

The oriented method of search is probably the best method to use, using a hoseline as the reference. The searchers should move laterally twenty crawls, either to the left or the right. Once at full extension, each moves as quickly as possible back to the hose and the oriented man. The searchers then move three or four feet back toward the door and repeat the process, doing so until they arrive back at the entrance. If the area is large and staffing permits, more than one search hose may be deployed. The searchers must resist pushing chairs in front of them or throwing them around. If they do, they'll only bump into them, again and again.

Again, significant search problems could be expected at a fire in this occupancy. Floor plans with maze-like cubicles and non-fixed tables and chairs are among the problems at this occupancy type. *Credit Coleman*

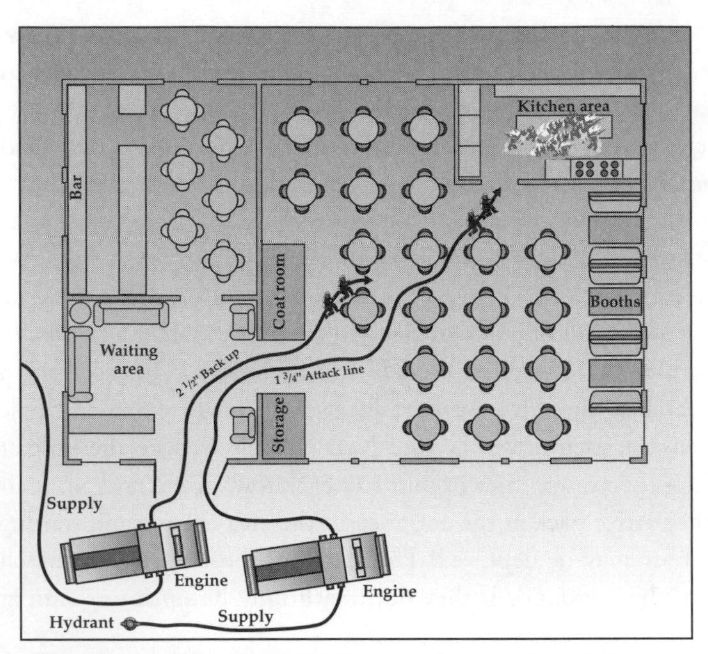

Backup should position itself approximately 2/3 of the way between the nozzle and the point of entrance.

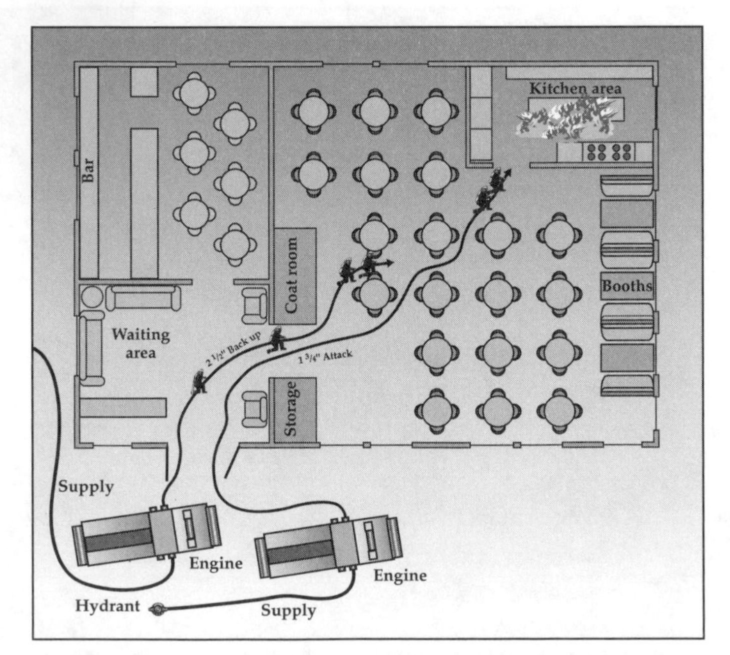

With long stretches, backup may want to position firefighters along the backup line to "stretch" the view of interior conditions.

Remember what is in your immediate area, and be happy that you ran into it again on your way back to the hose. If you hit a round table with movable chairs around it, then lay the first chair you hit on the ground. Then, work your way around the table until you come back to that chair. Leave the chair on the floor and move to the next table. Repeat the process until it's time to work your way back, using the toppled chairs as a guide.

By establishing a rescue crew, you immediately increase the effectiveness of the operation. It's inefficient if your searchers have to leave the search area every time they locate a victim. Post a rescue group to respond whenever someone is found. The oriented man is the one who should call them in. The team then follows the line up to the oriented man, where the searchers pass off the victim. The searchers can then rejoin the pattern without duplication of effort, and the rescuers can take the victim out along the line to waiting EMS personnel.

Consider ventilation to be a dire necessity in a restaurant fire. If the strategy is to defend in place, ventilation must be rapid and thorough. If the fire is on the lower floor of a multi-story building, then horizontal ventilation using fans will work best. If the fire is in the front of the building, take the fan to the rear. If the fire's in the rear, take the fan to the front. The same rules apply if PPV happens to be your method of choice. Positive pressure will certainly move a lot of air, meaning it will also clean out a lot of smoke.

As simple as it looks from the outside, significant fire problems can be encountered in these occupancies. Several firefighters have lost their lives fighting fires in this Restaurant type. *Credit Coleman*

RIT teams will serve themselves and their brother and sister firefighters well if they pay attention to their observations during a 360 walk-around of buildings. I see several ingress points on each Division on side "C" of this older Restaurant. *Credit Coleman*

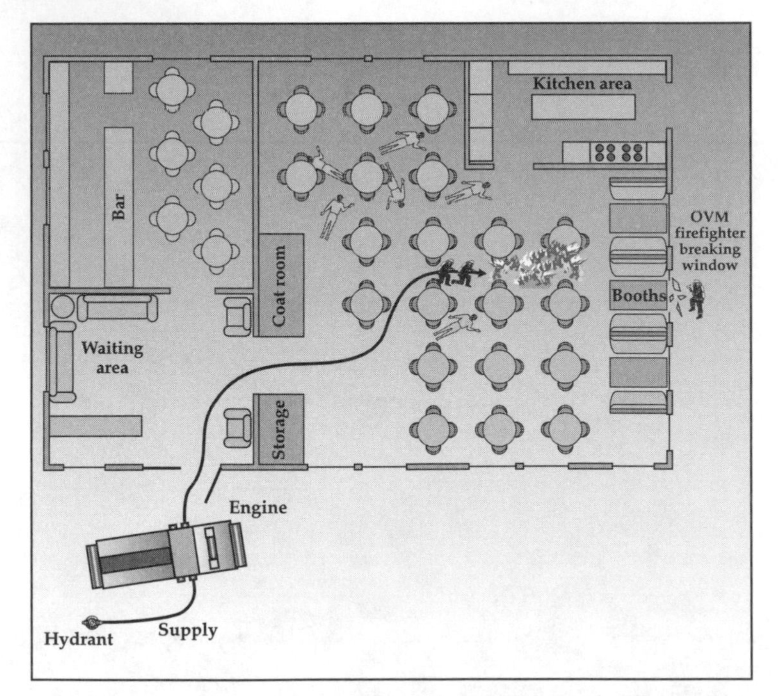

In extreme conditions, when staffing is low and probable life hazard is high, attack may delay tending to located victims in order to quickly attack and knock down the main body of fire. This action of passing victims may "do more to save more" than dropping the line and removing the first victim you find.

You may wish to use roof ventilation when the fire is on the top floor of the structure or when the smoke and heat can be diverted to where it will do no harm. I wouldn't hesitate to vent a roof over a working fire on a lower floor—just as long as the dominant vertical channel will funnel the smoke to an area where life safety isn't a concern. I would especially consider using this method if my crews weren't proficient in PPV techniques. Staffing constraints may force you to use PPV, but if staffing isn't a concern, then vertical ventilation through the roof may be the best and safest option.

Backup lines are mandatory. Restaurants often have side rooms and cozy nooks and crannies. Once crews are inside these spaces, fire can flare up, wrap around them, roll over, or drop down on them. Exactly when a backup line is assigned, however, depends on the size of the occupancy, the nature of the fire, whether savable victims are inside, and, of course, manpower.

The backup crew should position its line more or less according to the same parameters as are used in a residential fire. If the fire is on the first floor, then the nozzle of the backup line should be about two-thirds of the way between the point of entry and the attack team. The backup line should be farther in if there are many little rooms involved; however, the attack crew still needs room to work. The officer assigned to backup should occasionally work his way up the attack line and check on the status of the attack crew. Likewise, two members of the backup crew should occasionally move back to the entrance to ensure that the way out is clear. As in a lumberyard fire, the officer may feel compelled to ask for additional members to be placed every fifty feet along the line.

If the fire is on an upper floor, then the backup line should be pulled up the stairs, and there should be plenty of slack. If a single crew is acting as backup for an attack team on the second floor, and if each floor is 75 X 75 feet square, then the length of the backup line should be the distance to the interior stairs, plus the length of the stairway, plus another 75 feet for the second floor. Again, position the backup nozzle behind the attack nozzle, approximately two-thirds of the distance beyond the threshold to the second floor.

A RIT can really have its hands full at a working fire in a restaurant. Fires in these occupancies are often underestimated for their potential. In many of these establishments, the interior decor is largely a matter of set design, and the construction may not be all that it appears. A good size-up is certainly in order, including a 360 walkaround. Egress is a special concern, and you should always play the what-ifs of roof collapse. If the roof comes down, the interior walls will provide the most protection inside. Exterior doorways may or may not withstand the fall. If the door is located in a glassed-in foyer, as many of them are in these occupancies, don't expect it to be a safe haven from collapse. A side or rear doorway in a CMU wall will provide the best protection, and even this can't be guaranteed. If the roof caves in, remember that the exterior walls can usually be breached easily.

Know the layout of the dining area, as well as where the kitchen, storage, bathrooms, and maintenance areas are. Small rooms have compact walls and closer corners, and these characteristics may be enough to thwart a roof collapse. The members of the RIT may have to enter through these areas when deploying to rescue a trapped firefighter. If the restaurant is large and has many rooms and wings, then the members of the RIT should consider going in with another hoseline crew to back them up. Never should RIT personnel have to handle a hoseline themselves, however. If you don't have help, get it. Pull out a nonessential crew and give those members a backup line.

Scenario

A fire breaks out in a restaurant on the outskirts of the city. The building is 140 X 60 and one story tall. The exterior load-bearing walls are of CMU covered with plaster for a stucco look. Inside, the building has wood-frame partition walls. The roof is flat and held up by steel-bar joists. There's a newer membrane covering a metal deck. The entire structure sits on a slab.

The main dining room covers approximately 5,400 square feet and has a capacity of four hundred, calculated at fifteen square feet per seated person, plus so many standees at the bar. There are approximately seventy tables, with either four or six chairs each. There's also a bricked-in area used for outside dining during the summer. One of the exits dumps into this area. Cooking, storage, and bathrooms comprise the remaining floor area of about 2,700 square feet.

This Restaurant, which serves as the fire building for this chapter, poses significant access and search problems. *Credit Coleman*

There's a built-in dry-chem fire suppression system protecting the grill in the kitchen, but the building proper is only partially sprinklered with an automatic system fed from a domestic supply. Essentially, the sprinkler is protecting the kitchen, and the system hasn't been tested since the restaurant opened.

The fire occurs on a Saturday just before noon. A waiter catching a fast cigarette accidentally and unknowingly ignites a mountain of paper towels stored near an open delivery door on the side of the building.

Small-Department Response

Life safety is the primary concern of the IC, but the key to fighting a fire in this type of occupancy is to flow water and ventilate aggressively. This is no time to worry about pushing fire throughout the building. This is the time to worry about clearing the building of smoke as soon as possible. Positive pressure will accomplish this.

The easiest fires to fight are those where the flames are showing on arrival. The worst are those where only smoke is venting. With fire showing, you have a good clue as to where to take the line. With only smoke, it may be hard to ascertain where the problem is, how big it is, and where it's heading. Given a small department and lots of smoke, it may be best to haul in a 1 3/4-inch line at first, since this will provide the mobility necessary to get to the fire as soon as possible.

An outside vent man from the first-in truck will give the fire a place to go, once its location has been determined. It isn't always prudent for the OVM to take windows and doors prematurely, since doing so might accelerate the fire or draw it behind the interior crews. The OVM should wait for instructions from the attack crew as to where they want to push the fire.

As per the usual protocols, if the interior crews must operate on multiple floors or over large areas, then a backup line manned by at least two firefighters will likely be in order. A RIT should be established as soon as staffing permits. When assigning crews, the IC should also use the workers of the restaurant to help get a head count of the employees and as many patrons as possible.

The secondary assignments are to complete the primary search, overhaul the fire area, determine the cause, initiate the secondary search, and salvage property. The health department or its local equivalent must be called in to determine what food items can and cannot be saved for later use.

Medium-size and Large-Department Response

The first-in engine takes on the attack, and the second-in engine assumes backup. For reasons of speed and mobility, the officer of the first-in engine should likely pull a 1 3/4-inch line for the reasons cited above. The second-in engine deploys a larger line however. This should go in dry and be charged after it's inside. This line should be relatively stationary. Still,

A view of the front of the Chi-Chi's Restaurant that makes this chapter's case study.

A rear view of the case study. Note the fire venting from the rear door. Lines taken in from the front should help in getting a line between the fire and savable victims.

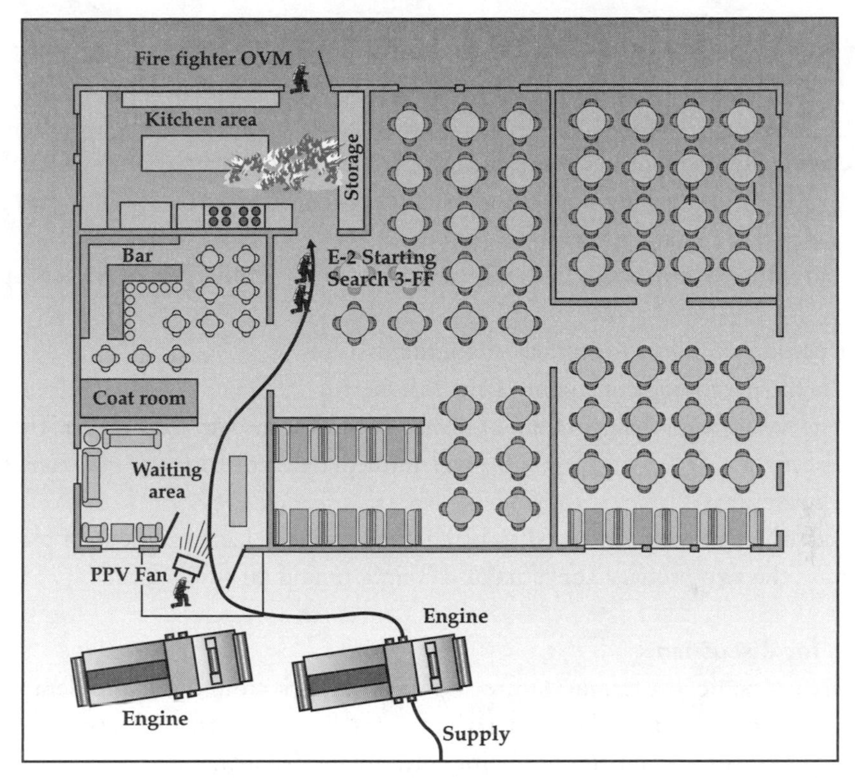

In a Small-Department response, initial efforts are to quickly knock down the fire and vent the structure.

four firefighters should be assigned to it in the event that it must be moved in quickly to back up the leading team. In a typical scenario, the IC should assign the squad to search operations, and the truck should split its crew to establish a rescue group and start PPV. As additional units arrive, the IC should assign a RIT, and more personnel should be allocated to search, rescue, and ventilation. After aggressive searches have been completed, the crews should take a break. Rotate firefighters in on mop-up lines while the initial crews are being checked at rehab. Concurrent to this and before overhaul, the investigator should look over the scene to determine the area of origin and the cause. Once these have been determined, and samples and photos have been taken, the crews can then overhaul the fireground.

Chapter Fourteen Questions

1. _______________ and ______________ are the two main problems in restaurant fires.
2. Panic generally does not become a factor in a restaurant fire. True or False
3. ____________ is the only way to understand how construction will factor into your ability to fight a fire in any restaurant.
4. If at any fire, you're torn between attack, search, ventilation or other operations, _____________.
5. When setting up initial fire attack, place the first line _______________.
6. What is the overriding constraint of the searcher?
7. Searchers should push chairs in front of them to best clear the way for a search. True or False
8. Establishing a __________________ will immediately enhance the effectiveness of the search operation.
9. What are the two areas that need to be checked after the backup line is in place?
10. What are the two primary concerns of RIT in a restaurant fire?

Questions for discussion

a) Discuss panic in a restaurant fire and ways to work around this problem.
b) The author discusses "level of training" for fire crews as it affects loss of life at a restaurant fire. Evaluate this factor within your department.
c) Explain the oriented method of search in a restaurant fire.

Chapter Fifteen

Fighting Fires in Churches

When a fire breaks out in a church or temple, much of the tradition of a community may be at stake. Even if a modern house of worship can be replaced, the memories of an incalculable number of ceremonies will forever be tarnished, and the loss of a grand, old church is akin to losing a page of history.

Many communities have experienced this loss. From the turn of the twentieth century up through the 1950s and '60s, many places of worship were destroyed. For some reason, there seems to have been a decline in these types of incidents. The reasons may be early detection, automated suppression systems, better training, and a variety of other factors. The Toledo chronicles list several major church fires. In my career, I can recall only one: Old St. Patrick's. In the late 70s, the steeple was struck by lightning during a summer storm. Crews battled the flames in tight, wood-laden areas, but they held the fire to the steeple with only associated water damage to the organ and choir area.

For the purposes of this text, there are essentially three types of church in use today. The first is the old, traditional, gothic variety, generally of ordinary construction. The second I will categorize as being modern traditional. These have been built since the 1950s, and they're more fireproof in nature. They tend to be smaller than their older counterparts, and although the basic layout of the nave and chancel may be similar, they generally provide an easier ground for fire operations. Finally, there are the conversions. Many buildings now in use as houses of worship were originally designed for some other purpose. Architecturally, the original structures can range from wood-frame storefronts to storage facilities to brick schools to strip malls.

In the old, gothic, cruciform variety, ventilation is far and away the biggest problem during a fire. The rooflines in these buildings can be beyond the reach of aerials, and even if the roof is accessible, the pitch of it may be more than 45 degrees. Slate roofs are the norm.

An old traditional Church poses significant fire problems. Ventilation tops the list at this structure. *Credit Coleman*

This modern traditional Church will pose some of the same fire problems as its older counterparts. Roof access will not be a concern however. *Credit Coleman*

An old traditional Church. This Church roof can not be reached by this city's tallest aerial. This will have a significant impact on ventilation options. *Credit Coleman*

Topside ventilation may be difficult to impossible. There may be a rose window way up high on either end, but think carefully before you take out any stained glass. No matter what your overall opinion of positive pressure, experts on both sides of the issue tend to agree that a church isn't the place for PPV if fire has entered the nave.

Inside, the height of the roof means that the ceiling area is also far off the ground, well beyond the reach of hooks. Hydraulic overhaul of the rafters may be the only option. Smooth-bore nozzles at high pressures will help you knock out hot spots from below, but this will result in excessive water damage from the heavy runoff.

There's so much in these structures that's irreplaceable, to flow water needlessly is virtually a sin. Many of these buildings have numerous void spaces, too, especially if the interior, nonbearing walls have been finished off with lath and plaster. Flames can enter the space behind these walls and run to the ceiling. If the fire breaks out in the basement, there may be slim hope for saving the building.

Modern churches may also have drastic roof angles, but the rooflines are usually closer to the ground and there may be better access, especially if the church was erected along a main thoroughfare. In my community, many of the modern traditionals have parking lots in the rear or along one side, whereas the older churches tend to be in congested downtown areas, tightly surrounded by various exposures. Some modern churches have very ornate, geometric roofs that may be difficult to ventilate. The covering may be of slate, tile, metal, or wood shingle. Suspect truss assemblies and all of their associated problems. The span of the rafters, usually of solid or laminated timbers, can be quite wide. They'll probably be tied to the walls on steel columns, with decorative wood or plaster trim to resemble corbels or imposts. Modern churches also tend to have more firestopping and other safety features built into the overall design. Steel will likely prevail in the exterior, load-bearing walls, and the piers and buttresses may be more decorative than functional.

Fires involving Churches should almost always require a 2 1/2" line as the initial attack line.

This Church was once a single family residence. What would you envision the floor plan to be like? Know your buildings and pre-plan. *Credit Coleman*

Predictably, the converted house of worship is the architectural wild card, since it may suffer from any of the deficiencies common to any other renovated structure. Code violations are often overlooked out of sympathy for smaller, poorer congregations. Older churches usually have fixed seating, which is an asset during primary searches. In a converted church or temple, however, firefighters may have to grope their way through a room littered with movable fold-up chairs. Many precious moments can be lost in the effort to maintain patterns and stay oriented in such a field.

Early collapse isn't usually a concern in old-style gothic and modern traditional churches. The roof assemblies can come down, but precipitous collapses are rare, and when they do occur, they tend to give plenty of warning. These structures are built to be voluminous, and there's always a direct relationship between the massiveness of a building and its fire-resistance. These statements do not apply to conversions, however.

I make no claims to be a historian, but I know of no large loss-of-life fires in churches—defining a large loss-of-life fire as one that claims, say, more than five individuals. Rolling up in today's apparatus, I wouldn't anticipate a large life hazard in a reported church fire. We recently responded to a second alarm in Toledo's Rosary Cathedral. This large, old-type traditional gothic church had a small basement fire that started at approximately 1830 hours on

An old traditional Church in a downtown setting presents significant exposure concerns. *Credit Coleman*

This is a photo of the exterior load-bearing wall of an old traditional Church. *Credit Coleman*

This is a photo of the exterior load-bearing wall of a modern traditional Church. There is considerable less "mass" than is present in the old traditional Churches. As a rule, the less the mass, the less fire resistiveness. *Credit Coleman*

a weekday night. Church services and social events were taking place, but there were only a few dozen people in the church at the time. The fire was quickly extinguished by a single 1 3/4-inch line. Ventilation, however, proved to be an extensive operation. In fact, we used several PPV fans and special-called a mutual-aid department for their hovercraft to help blow away the smoke. There were no injuries; just smoke in all the high recesses of the church.

Strategically, you must make every effort to maintain interior operations if you expect services ever to take place in that house of worship again. It's a general rule that you must get inside and fight the fire directly if you plan to save the structure. It's been stated above that old-style and modern churches are resistant to collapse. They also provide many places from which to operate in relative safely. Arches and other well-supported portals come to mind. Make no mistake, if fire is attacking the structure and time has ticked past a time frame of ten to twenty minutes or so, you as the IC must consider withdrawal, and soon. Even with the biggest and the best and the oldest, you should always establish collapse zones at well-advanced fires, just in case. You may even want to have more than one safety officer at a working church fire.

At a Church fire when high heat is present at ceiling level, don't hesitate to throw water at the upper portions of the ceiling.

If you can't ventilate successfully, your ability to save the upper portions of the structure, and thus the entire structure itself, will be questionable. If you absolutely can't open up the roof, apply as much water as necessary to cool the interior and simultaneously extinguish the fire. Go big and go fast. Yes, you'll cause a lot of water damage, but the alternative might be eventual collapse. Big departments can pull multiple 2 1/2-inch lines. Midsize departments can probably do the same. Small departments may need to set two deck guns or monitor nozzles in the door and flow 1,000 to 1,500gpm at the fire. I know of no other alternative. The game is to put out the fire. In my opinion, there's no excuse for any line smaller than 2 1/2-inches to be pulled when there's evidence of a fire in a church on arrival of the first unit. Suppression operations should have three qualities: speed, strength, and simplicity. Speed counts, so commit your first engine to the front of the building. Pull preconnected lines as much as possible. Some stretches, of course, will exceed the traditional lengths of preconnect hoselays. Take in the lines quickly and from the most appropriate area. What is the most appropriate area? One that will cut off the fire to the smallest area possible. If the fire is in the front, then the line should come through the rear or a side door. If the fire is venting from a side window, do not shoot water at the window. Move to a door that will allow access to an area where you can stop the progression and push the remaining fire out of the hole.

An old store-front that has been converted into a Church. Expect movable seats in this Church. *Credit Coleman*

Normally, search won't be a prime concern at a working fire in a church. Given warning, most of the congregation will be able to self-evacuate. Still, flash fires lit by incendiary devices and incidents that occur after hours may mandate search, especially for live-in clergy. In old-style and modern traditional churches, the prime places to search include the chancel, vestiary, nave, foyer, choir loft, basement, and the classrooms and meeting rooms of adjoining structures. An oriented man working the center aisle of the church can conceivably control four or even six searchers—half to his left and half to his right.

Having checked the center rows of pews, they'll then need to search the rows along the side aisles, again working toward the rear of the church. The chancel can be swept by one or two searchers before they move to the side aisles. It's best to use a hoseline for orientation when searching the basement. Schoolrooms can be undertaken using the oriented method, with the oriented man working off the hallway. Deploy one searcher per room, with a maximum of three per oriented man. These rooms can be large, and maintaining contact may be difficult.

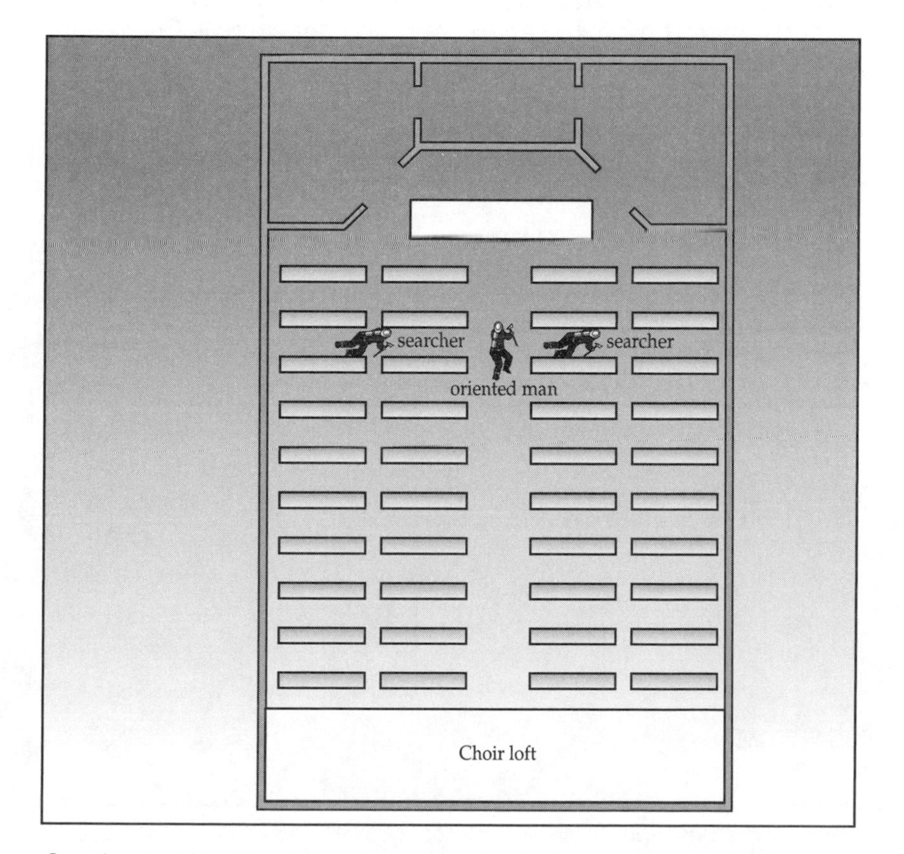

Searches in old traditional Churches can be done by following the layout of the pews.

The interior of an old traditional Church. These pews should aid in search efforts. One firefighter can move quickly between the pews. *Credit Coleman*

Many old traditional Churches have massive roofs that are difficult if not impossible to ventilate. If roof ventilation is contemplated, aerials will prove the safest means of access to the roof. *Credit Coleman*

Search in a converted church, of course, can be problematic, and no text could even begin to detail all the myriad hindrances that you might encounter. If the area is large and you're not exactly sure what's in there, take in a 2 1/2-inch hose with you as your line of orientation.

Backup lines are essential at fires in churches. Generally, if the fire is in the church itself, then one large-caliber backup line should suffice. The nave usually only requires one crew with a 2 1/2-inch line. If the fire is in the basement, then several lines may be required. The below grade areas may be huge and labyrinthine. Crews can easily become disoriented and cut off from their route of escape. If you take an attack line downstairs, it should be long enough to reach to the far end of the basement. As always, you should position the nozzle of the back-up line two-thirds of the way between the stairway and the attack crew.

There are no specific rules to follow for salvage, but let common sense prevail. The greatest concentration of precious items in a church or temple will be around the altar. Try to remove statues, crucifixes, chalices, menorahs, and other objects of value to a safe area within the building, if possible, and avoid taking them outside, where vandals or the elements may do their worst. If staffing or time constraints don't allow you to remove them, then cover them

RIT would be looking for access and construction features in their 360 of this old traditional Church. *Credit Coleman*

with tarps. The vestiary is another important area for salvage. If you must take them outside and staffing is low, use a police officer to guard the objects.

The sheer size of some of these structures will top the list of concerns for the RIT crew as they make their 360-degree walkaround. Visits to the command post to meet with the IC and the operations officer are a must, and the RIT officer will want to ensure that the department's accountability system is accurate to the minute. If activated, it may be best if the members of the RIT are accompanied by a safety officer or some other experienced veteran. Church fires can change character quite suddenly. They can flare up rapidly and evolve through a course of events leading to flashover or collapse. Having one firefighter who can keep watch on the conditions inside can make all the difference in the world to a team focused on the complexities of rescue.

Scenario

A fire occurs in an old-style church in the downtown area of the city. Built in the late 1890s, it measures 60 X 130 feet. The exterior load-bearing walls are of masonry, with buttresses to help support the weight of the walls and roof. The huge wooden arches that carry the roof are laid into the buttresses. The roof is covered with slate; inside, there's lath and plas-

The simulation Church. This is a photo of sides "A and "B". *Credit Coleman*

ter attached to wooden planks. The interior walls are also of lath and plaster. There is a full basement, and the twin stairways leading to it are located in the vestibule.

The fire occurs on a warm Saturday evening, shortly after mass. Only a handful of people are still in the church. Fans were used during the mass to cool the parishioners, and now there is an increasing smell of something electrical burning. As the priest and a layperson look for the source of the odor, flames suddenly break out at floor level near the altar.

This scenario begins like many major fires—as a small, innocuous run. Not a lot is showing from the outside. In describing their impressions of the opening stages of such incidents, officers and firefighters often comment that "everyone and everything seemed to be moving in slow motion."

Small-Department Response

As soon as the initial officer enters the church, he realizes the potential and asks for more help. Some would call this a precautionary second alarm, but I would simply term it smart. I can think of a multitude of reasons to call for a second alarm in this scenario. I can't think of one reason not to.

The first line to go in should be a 2 1/2-inch, since nothing less is appropriate for the potential of fire in a church. At the time of arrival, the extent of the fire and its origin are

The fire in this chapter's case study as seen upon arrival of the first unit.

unknown. With three-person engines, having the manpower to handle the long lengths of 2 1/2-inch hose will be a concern. Two crews will be used on the initial line, meaning four members for attack. The line from the second engine is also a 2 1/2-inch. The IC commits all of these resources into the basement. One of the hoses will advance on the fire, and the other will be used for backup. In this evolution, the backup line is charged first and pulled by the four members along with the dry attack line. The IC's thought is that one member can maintain the backup line in a stationary position behind the attack crew, and the other three members can stretch the attack line as close as possible to the fire before charging it.

Until the second-alarm units arrive, the IC only has a few more members to assign. In this case, he tells the truck crew to pull a line to hit any fire that vents up to the first floor around the altar. With these assignments, the commander has a suppression force operating in the basement and one to hold back the fire and keep it from overtaking the church itself.

Given that these crews can wage a successful battle against the flames, ventilation soon becomes the priority. The second-alarm truck uses PPV to clear the church of smoke. Knowing the dangers of PPV, the IC orders the attack crew to remain in place while the fan is operating so as to monitor any flare-ups that occur. The incident commander should also assign a RIT as soon as staffing permits.

The fire in this case study as seen 10 minutes into the fire.

If there are no reports of persons missing, salvage and overhaul will be the most important secondary assignments. A fire in a church is one incident in which salvage actually takes precedence. Even so, as these operations get underway, the IC assigns two firefighters to conduct a search. As soon as salvage operations have been completed and the cause determined, overhaul can be finished and the crews can begin to return to service.

Medium-size and Large-Department Response

One should never be complacent about a church fire. Therefore, a commander from a larger department would be equally prudent in issuing a second alarm at the first sign of fire. The initial deployment is the same: two sizable hoselines into the basement, one for attack and the other for backup. With more resources at his disposal, the IC should send additional lines into the church itself to handle any fire venting from below. He could also deploy an outside vent man and a forcible entry team to the rear of the church, just in case the fire has taken a devious turn somewhere. Given that the offensive operation is successful, the secondary assignments would be the same, and the IC would be able to establish a RIT and mount search operations earlier. Besides sheer numbers, this is the value of a larger response: that more operations can be mounted concurrently. It's likely that the crew of the first-in second-alarm truck would be chosen for the RIT. After the fire has been knocked down, the crews of the second alarm should be used for salvage while the first-alarm crews conduct overhaul.

The fire in this case study as seen 10 minutes into the fire.

Chapter Fifteen Questions

1. Topside ventilation is normally difficult to impossible in a working church fire. True or False
2. When overhauling the underside of a ceiling in an old church, ____________ may be your only option.
3. Older churches usually have _______ seating.
4. Converted churches generally have _______________ seating.
5. Early collapse _______________ usually a concern on old-gothic churches.
6. The author gives from _____ to _____ minute time frame before collapse can be anticipated in an old-gothic church.
7. According to the author, there is no excuse to pull any line smaller than a _____ with any evidence of a fire in a church.
8. In older gothic churches, what feature will aid in search efforts?
9. Salvage efforts can be very labor intensive at a church fire. True or False
10. How should searches be conducted in converted churches?

Questions for discussion

a) Converted churches pose a wide variety of problems to firefighters. Survey your city and come up with what you consider the three most uncommon original occupancies that are now converted churches.
b) Discuss the pros and cons of PPV in an old gothic church.
c) Discuss the value of church artifacts versus the staffing required to protect it. Should additional alarms be sounded in fires in these occupancies specifically to protect religious articles or is this a waste of staffing and reduced fire protection in other areas?

Chapter Sixteen

Fighting Fires in Nursing Homes

Year after year, some of the highest-cost fires in terms of loss of life occur in nursing homes. These occupancies can house hundreds of elderly, often physically handicapped individuals, living twenty-four hours a day under the care of ever-dwindling staff. Up until a few years ago, construction and maintenance were key problems in nursing homes. Many of them weren't kept up and were usually in need of repair. If there were sprinkler systems in place, they were often neglected. Today, backed by more stringent codes, certifying agencies keep pretty close tabs on these occupancies. All nursing homes must now be equipped with properly functioning automatic sprinklers, and they must also be inspected by fire prevention professionals on a regular basis. There are wonderful homes in existence today that are well maintained and whose owners have the best interests of the residents in mind. Unfortunately, there are still many homes being operated solely for the money.

Far and away, the biggest problem associated with these occupancies is life safety. Try to imagine a building full of octogenarians trying to amble down stairways and corridors to escape the smoke and flames. Like other institutions, the floor plans of these occupancies are often confusing, especially those that have been transformed into nursing homes by conversion. Like hospitals, the rooms are small and numerous, and each building is bound to have its own idiosyncrasies. Don't expect to respond to a fire in a nursing home and find dozens of attendants evacuating the elderly and the impaired. The skeleton crews that run these places nowadays will probably best be able to serve you by providing an accurate accounting of who's still inside.

We will concern ourselves with four construction styles prevalent throughout this industry: the one- and two-story wing-style nursing homes, the converted structures, low-rise nursing homes, and high-rise nursing homes. Each of these presents different fire concerns, as well as strategic and tactical considerations.

The front entrance to a one story "wing" type Nursing Home. *Credit Coleman*

A high-rise Nursing Home has advantages and disadvantages to its low-rise counterparts. The presence of sprinklers is a plus but the sheer number of residents present and access concerns would be considered negatives. *Credit Coleman*

The one-story wing-style nursing homes are normally of wood-frame or ordinary construction, with exterior load-bearing walls of wood, block, or occasionally brick. The interior load-bearing walls generally have wooden-stud or concrete walls. In wing-type versions, the wings are attached to a central hub, typically a commons area. Some of these buildings are laid out in a square, with a courtyard inside. The roofs generally follow a wood-rafter or truss design, with plywood or chipboard sheathing and asphalt shingles. Trusses, of course, are more common in the newer buildings, whereas older buildings tend to have conventional rafters. The floors are either of poured concrete or wood-joist assemblies. Some newer wing-style nursing homes may also have truss floor assemblies, of either parallel-chord wooden trusses or engineered I-beam joists. Over wood joists, expect plywood sheathing, and the majority of these will have rolled or square linoleum floors for ease of cleaning. If you find a concrete floor, there will be either a glaze finish or tile over the concrete.

Predictably, the nursing-home conversions vary widely in design. Many in my area have true balloon-frame exterior walls. Many have been rendered out of old, large brick homes. Some of the conversions have brick veneer over wood-frame construction, whether balloon or

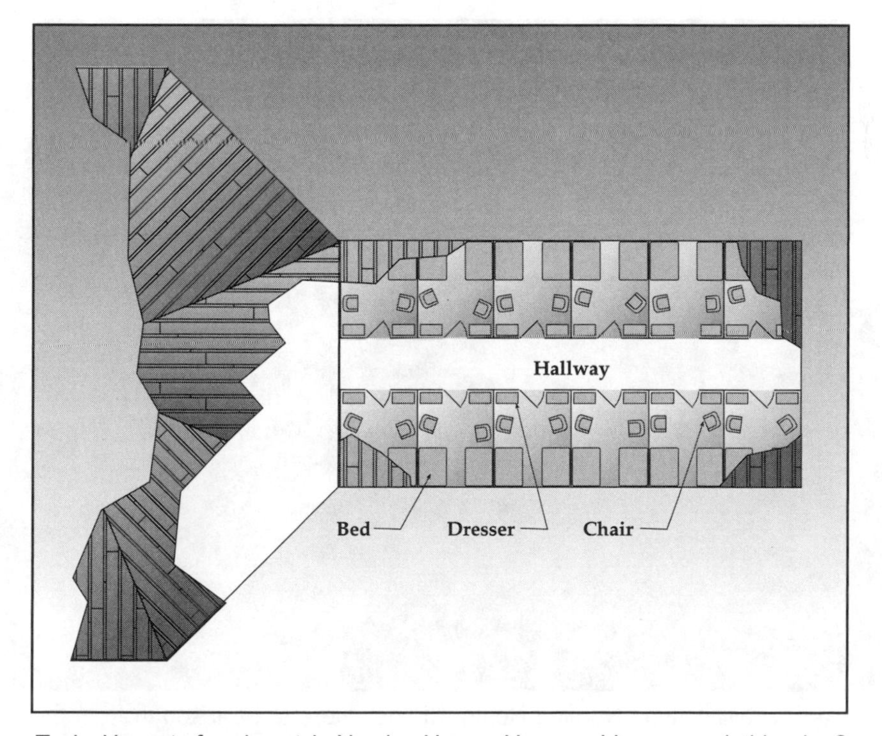

Typical layout of a wing-style Nursing Home. How would you search this wing?

platform. The interior walls generally consist of wood studs, plus drywall or lath and plaster. Wooden rafter assemblies predominate in converted nursing homes. These will have purlins or plywood sheathing. Asphalt shingles are most common, although you'll find both metal roofs and terra cotta tile. The floors are normally of wood-joist construction, with tile over either plywood or true hardwood.

Low-rise nursing homes tend to be of ordinary construction or a combination of steel and ordinary exterior load-bearing walls. There are some three-story wood-frame low-rise nursing homes. These are similar to garden apartments in structure but not layout. They have larger rooms connected to a major interior hallway, and in the majority of these buildings, all three floors are abovegrade. Most are in compliance with the Americans with Disabilities Act. The interior wall assemblies consist of wood studs or sometimes block. Newer low-rise nursing homes generally have wood-truss roofs with asphalt or perhaps tar and gravel. Some have roofs of poured concrete. Still others have steel-bar joists with metal-deck or built-up roofs. You'll also find rubber-membrane roof coverings out there. The floors of both low- and high-rise nursing homes are often supported by steel-bar joists, with Q decking plus a thin layer of poured concrete. Some have poured concrete supported by steel and columns. For sanitary reasons, carpeting is rare except in entrance areas. Most residential areas have tile or concrete.

The rear of a new "wing" type Nursing Home. If lines were to be taken in from this direction, would your pre-connects be sufficient? *Credit Coleman*

High-rise nursing homes are found all over the country. They can stand more than twenty-five stories tall and house hundreds of residents. Some of them have private bedrooms arranged along a main hallway. Others follow more of a traditional hospital motif, with wards that accommodate a number of residents. These structures generally have load-bearing walls of steel, concrete, or a combination of both. The interior walls can be of drywall on steel studs; otherwise, they may be of CMU. The roof will almost certainly have steel-bar joists topped off with a metal deck, built-up weatherproofing, or a concrete assembly. As in low-rise nursing homes, the floors are usually some conglomeration of steel and concrete. Taken as a whole, high-rise nursing homes tend to be very compartmentalized and can produce hot fires.

Regardless of the type of construction, the key concern is any given building's ability to spread or contain fire by any means. In any building with a wing design, fire may travel along a given wing and should be stopped before it reaches the hub of the building. Even so, smoke

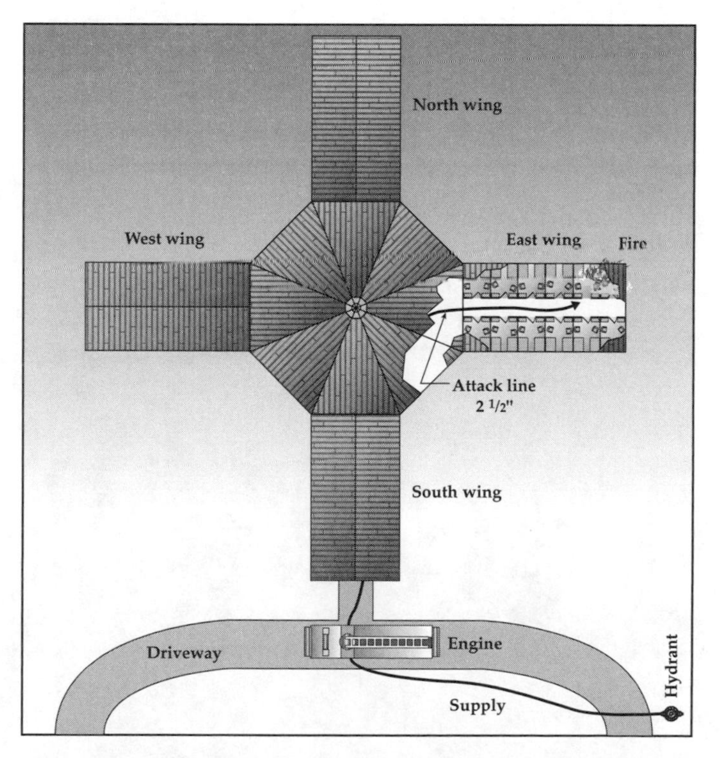

With a fire in a wing of a Nursing Home, initial lines should be taken in to get a nozzle between the most savable victims. In this fire it is justifiable to take the "long stretch" in order to protect the most victims.

can easily permeate and contaminate the entire structure. As a group, new low- and high-rise nursing homes are very effective in retarding the passage of fire. Built-in fire doors, fire-resistive members, and compartmentalization can help hold flames to their area of origin. Interior vents and duct systems present the main avenues for extension. Many of these systems have fire dampers or sprinklers built into these ducts. One source of worry in my community is the small rubbish fire in a shared trash chute. These chutes traverse all the floors and terminate in the basement or compactor area. Although built-in detection and protection systems may keep the fire confined to the dumpster, a smoky fire may spread its smoke throughout the building until responding crews can douse the flames.

As with other modified buildings, the greatest challenges are posed by conversions. These buildings were put up years ago and for wholly different purposes. In my community, most of them were originally large mansions or private homes. They have since been subdivided and expanded. Bedrooms have been converted into apartments housing two or more residents. Expect these buildings to react to fire the way others of their kind do. The good news about these mini institutions is that they house far fewer residents than other styles of nursing homes. There will likely be fewer than thirty residents. Looking for thirty individuals may be a chore for search crews, but it's certainly less of a problem than searching for three hundred in a bigger building.

The interior of a "wing" type Nursing Home is very similar to the layout of a hospital wing. *Credit Coleman*

Assessing the structure, the size of the fire, the number of residents, and the capability of your own department will reveal the true potential of an incident. Some of these structures will assist you in controlling the fire. Although feared by many firefighters, high-rises provide a good, safe platform from which to wage the battle. Still, these occupancies can produce large quantities of fire and get very hot as a result of compartmentalization. Most of the residents have difficulty caring for themselves under normal circumstances. Some are nonambulatory. Others have respiratory ailments that will be exacerbated by smoke. Others are mentally impaired and don't make good decisions. The behavioral component is one that you will never quite be able to predict, and it will play a large role in the final outcome. Remember, too, that a fire in a nursing home will draw a lot of attention from the press and the community as a whole—just one more reason to conduct thorough, responsible fact-finding inspections prior to an incident and to formulate a number of contingency plans to match whatever circumstances you might one day encounter.

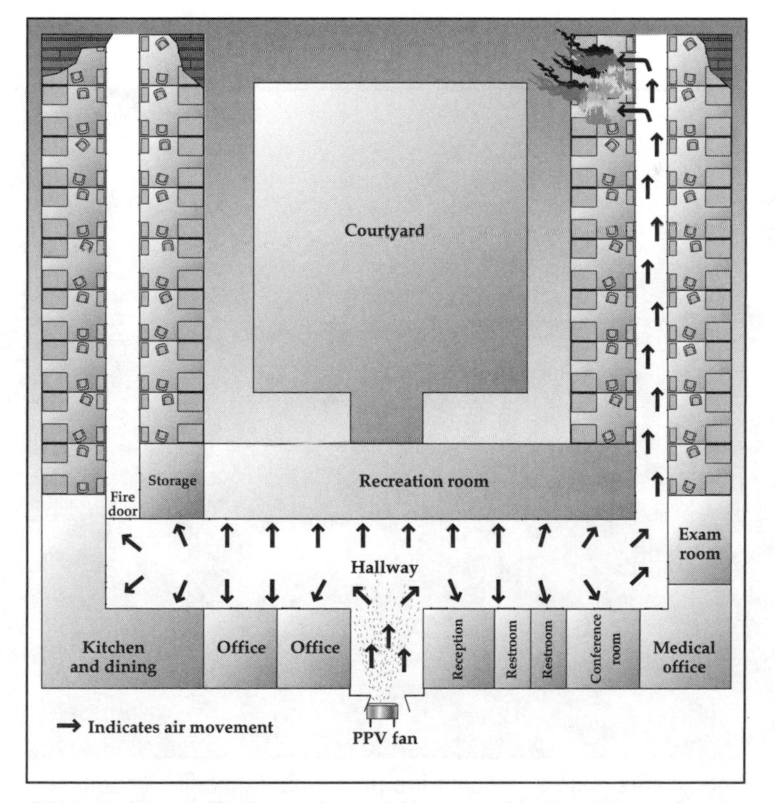

PPV can be an effective tool to quickly vent a Nursing Home after the fire has been knocked down and an exit area provided.

Given the life hazard, an incident commander's primary concern is manpower. Most departments can't respond with the number of members required to accomplish concurrently all of the operations necessary to alleviate the life safety issue. Most departments have to make do with what they have. The four principal tasks are to put out the fire, ventilate, search, and remove the residents. To be effective, search operations will require a minimum of a dozen members in a small environment. A large-scale nursing home will require far too many firefighters to count. Remembering the fifteen-minute rule, if you can't bring a victim out of an IDLN environment in fifteen minutes, you may as well start body recovery shortly after. Stated another way, there's no need to have rescue crews if they won't be able to bring anyone out within fifteen minutes. And the coroner will want to take pictures later. If you intend to perform rescue, you'll need a minimum of two members per crew. A rough rule of thumb is one rescue group per victim per every two to five minutes, or the time to drag out a victim and pass him to the medical unit. Ventilation will take anywhere from two members for PPV, up to twelve for roof cutting. Finally, attack will require a minimum of two crews, each consisting of three firefighters at minimum. As an average, the bare-bones number is

Converted Nursing Homes pose significant fire problems. Maze-like conditions, lack of built-in fire protection and construction concerns top the list. It is imperative that these occupancies are pre-planned. *Credit Coleman*

twenty-six. If you can't bring twenty-six members to the "dance," then Plan B may be in order. Defend in place. Take everything you have, put the fire out, then vent. You'll probably do more to save lives by following this prescription than you would by sending a dozen members to comb through a big, smoky building.

If searches are to be conducted, the oriented method will work well. A single oriented man can manage three or even four searchers. Most of these buildings are relatively easy to search because the living quarters are usually right off the hallways.

Attack operations aren't complicated in nursing homes, either. As a rule, the attack should commence from a point where you can get a line between the majority of the residents and the fire as quickly as possible. In a wing-type structure, the architecture can benefit us, since the wings are accessible from the hub. Lines that pass through this hub can be directed down any wing to cut off the fire. Some buildings have inordinately long wings and several sets of self-closing doors, so bringing the hose through the hub isn't always the best option. In a converted structure, the center stairs normally provide the best route upward. Except for fires in the vicinity of the main entrance, the front door is usually the best means of ingress. Fires in low- and high-rise buildings are generally fought using stairways and standpipes; otherwise, fire towers.

In any situation with a high life hazard, always take more water in with you. Pull two dry 2 1/2-inch lines, then charge them to flow at least 250gpm each. Go big and go early.

When it comes to ventilation, remember that we open up buildings for three reasons: one, to reduce the effects of contaminants on civilians; two, to reduce smoke damage; three, to make our job easier by increasing visibility. Of these reasons, the first is assuredly the most important. In a wing-type structure, we must ensure that any self-closing doors that will isolate a given wing are actually closed. Sometimes they're propped open during the daytime for the convenience of the employees. The most effective way to get rid of the smoke is to open the roof. This should immediately help to lift the smoke off of victims without blowing it around to other areas. This is especially true in one-story buildings or two-story structures where fire has entered the walls. A single hole in the roof and ceiling over the fire, if possible, can work wonders. A less favorable option, in my opinion, is positive pressure. If possible, the best place for a vent hole is in the fire room itself. Open a window. An open window in an unoccupied room is your next best choice. Your last best choice will be an open door at the end of the hall. If there's only a self-closing door at the hub and a door at the end, then place the fan at the hub and blow down the hall.

I realize that the manufacturers say that these fans shouldn't be operated inside a building, since they produce carbon monoxide. Trust me, the residents won't complain about exhaust fumes and a little extra CO if the air can be cleared that much sooner. If there aren't any PPV fans in your department's inventory, negative-pressure fans will also work. As a last resort, natural horizontal ventilation will work, but it's extremely slow.

High- and low-rise nursing homes require a different approach. Traditional ventilation using fire towers or stairways works well. Some of these stairways are pressurized, or we can pressurize them with PPV, also. As the air and smoke rises in the tower and past the fire floor, it will draw smoke from down the hall by means of the venturi effect.

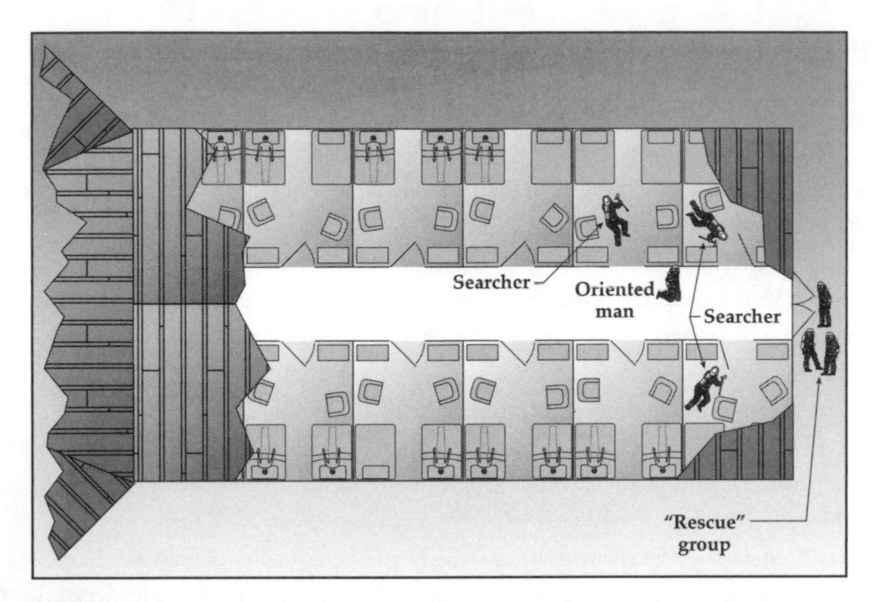

Searching a wing of a Nursing Home using the and utilizing a rescue group outside.

The rear of a "wing" type Nursing Home. Assess should not be a problem in this complex. *Credit Coleman*

Converted structures, of course, should be vented according to the type of structure on which they're based.

The stumbling block with respect to backup is the life safety issue balanced against the available manpower. If staffing is low, all forces on the first alarm must be committed to locating and extinguishing all visible fire. Next, the products of combustion must be eliminated from the interior. To use personnel for backup under these circumstances might be a waste of manpower. After all of the visible fire has been extinguished and the smoke has cleared, what would you like a backup crew to do?

In considering the most extreme circumstances, I can envision one or two cases in which backup teams might be assigned to protect the interior crews. If the attack crew must drive the flames back to the area of origin and there is a definite possibility that these members might be cut off by flare-ups as they proceed down the hall, then assigning a backup team of two members or so is justifiable. Managing minimal resources to the best effect requires tough decisions at times, but it's the ability to do so that defines a true leader. Such decisions never come easy, and next-day hindsight will often point out the preferred, often obvious course of action that should have been taken. If the loss was big enough, such hindsight can drive even the best officers and chiefs to early retirement. Since the risks of our industry are so high, it's vital that you honestly account for your capabilities, as well as your department's, during the pre-incident phases. Fight your fires under the expected conditions so that, when the worst happens, you'll already have half the battle won.

The same comments probably apply to a RIT. You may not have the manpower at the outset to establish one. You'll eventually need a RIT, however, so as additional units arrive, you should make this assignment as soon as possible.

The fundamental reason for using a defend-in-place strategy in a nursing home is a lack of manpower. Normally, when staffing is at a shortfall, we tend to put others above ourselves. We make do without. We throw caution to the wind and stretch ourselves well over the line. The heroic impulse runs strong throughout the fire service, but it's at these incidents, where we bust our guts to save lives, that we all too often become the victims. Whether by procedure, additional alarms, mutual aid, or some other means, the IC needs to get a RIT in place at a working fire in a nursing home. Interior crews can then afford to go the extra distance if they know that someone is there, watching over their backs.

After performing a 360 of the building, if activated, the members of the RIT will normally want to enter the building by the same route as the attack crew, from the unburned portion toward the burned portion. Aggressive operations will be going on inside. It may be that the interior crews are knocking down fire and pushing forward. Flare-ups may threaten them from behind. This may also be the one time that I would advocate using a smaller 1 3/4-inch line for backup rather than the 2 1/2-inch line that I ordinarily recommend for commercial occupancies.

Many nursing homes have an abundance of windows, meaning all the more avenues of egress for firefighters. If you need to take out windows, take them in full. Leave no wood, glass, or metal standing. The short stature of many of these buildings also facilitates rescue. During the walkaround, look for the easiest ways in and out, and study the configuration of the windows. Are there two windows per room or one? Does it seem that each room has a window, or are they spaced too far apart for that?

Events happen quickly at a working fire in a nursing home, and accounting for all the personnel inside will be difficult at best. Try to get a RIT member stationed near the command post to monitor the accountability board. The rest of the crew can consider the what-if scenarios.

You must also account for those who come out of the building; i.e., the residents. Accounting for the rescued is relatively easy. For the most part, there won't be too many of them, and they'll normally be transported to a medical facility for additional treatment. The EMS provider usually has the time to get the names of these people. Accounting for the evacuated is another story, however. If you don't assign some competent party to corral and control the evacuees, you'll almost always have to face some degree of embarrassment. Some loved one will almost certainly walk up to you at some point during the incident and ask, "Where is my Aunt Tilly?" Most chiefs don't like to be forced into saying, "I don't know!" Most responsible chiefs would rather say something to the effect of, "All the evacuees have been taken to the church across the street. Lieutenant Jones should be able to help you find her."

And just who is an appropriate "competent party" to oversee this task? Generally, I would recommend that you use a firefighter, since he will signify authority at a fire scene. He is also trained to have an eye for EMS concerns, and he is intimately familiar with the hierarchy of the responding forces. In a small department, you might have to enlist a police officer, an employee of the nursing home, a clergyman, or some other qualified candidate. Be selective in your choice. Have that person obtain from the evacuees, as quickly as possible, all of their names, room numbers, and any other pertinent information, including information about prescriptions and medications. The evacuees should be allowed to phone their loved ones, especially those relatives who may live in the area and see reports of the fire on the local news. Out-of-town relatives can be contacted later. Finally, it wouldn't hurt to have someone take the vital signs of all those who have been evacuated. A private ambulance crew or mutual-aid EMS team can assist in this.

SCENARIO

This is the nightmare scenario. Nothing about the building, except for a skylight on the top floor of one wing, favors the firefighter. If you're lucky, there won't be anything even remotely like this occupancy in your jurisdiction. In all likelihood, there is.

This nursing home was built around 1935. It is of ordinary construction and has exterior load-bearing walls of brick. The building stands two stories tall and has a full basement. There are wooden floor joists that span from wall to wall. The floor is finished with wooden

An elevated view of the fire simulation building. Note the roof layout. Would you put crews on this roof to vent? *Credit Bob Oliphant.*

Side "B" of the fire simulation building. *Credit Bob Oliphant.*

planks and covered with linoleum. The roof has wooden rafters, finished to replicate a flat roof. The ends are hip assemblies. The rafters are covered with purlins. On top of the purlins are several layers of tar paper, plus a rubber membrane.

The building is laid out in two wings, one in the front and one in the rear. Between the two wings is a passageway, with a fire door separating each wing. There's also a storage closet in the center section on both floors. Each wing has approximately 1,500 square feet of floor space on each level. Each individual room measures, on average, 12 X 13.

There are plans to bring the home into compliance with the Americans with Disabilities Act, but this hasn't been done yet. The stairways, two in number, are located at the ends of the building. They are open up to the second floor and down to the basement. A fire door separates each stairway from the living areas. Each floor in each wing has four rooms on each side of the hall, with three beds per room and two windows, making for a total of ninety-six residents throughout the building. The home is at capacity at the time of the fire.

For the most part, the ambulatory residents occupy the second floor, and nonambulatory residents live on the first floor. There is a driveway around the building, and there's more parking in the rear. Fire apparatus can maneuver around the building as long as no cars are parked in the front or rear corners of the drive. The fire department has warned the owners that, if they allow parking in the first spots in the front and rear, fire apparatus can't get through. Several warnings and citations have been issued.

There are no residential units on the lowest level. The basement contains a kitchen, a dining area for the ambulatory residents, laundry rooms, a game room, and a recreation area.

Having never varied in purpose since the time it was built, the home lacks built-in fire protection. There is a central-station fire alarm. There are twelve daytime staff members, and at night the staff drops down to three. The daytime shift begins at 0630 hours.

The fire occurs at 0600 hours on a Sunday morning. It starts in the first room on the first floor in the front of the building.

Small-Department Response

The first concern is to get a line of adequate size between the fire and as many residents as possible. In this scenario, the fire is in a bad location. Taking in a line from the front won't get it between the victims and the fire. In some extreme cases, where there is no other option, this type of placement may be all that's available. In such a case, a smooth-bore nozzle, directed first at the ceiling to darken the fire, is a must. If no smooth-bore nozzles are available, then you should set your combination nozzle on straight stream. This will push the fire down the hall the least and generate the least amount of steam.

For the scenario as described, the officer creates a door out of a window. He takes a risk and chooses a window beyond where he estimates the fire to be. He and a crew member from the first engine take in a 2 1/2-inch line from there and advance on the flames, moving

The fire as seen by the first arriving unit. This is side "A" and "B" of the Nursing Home.

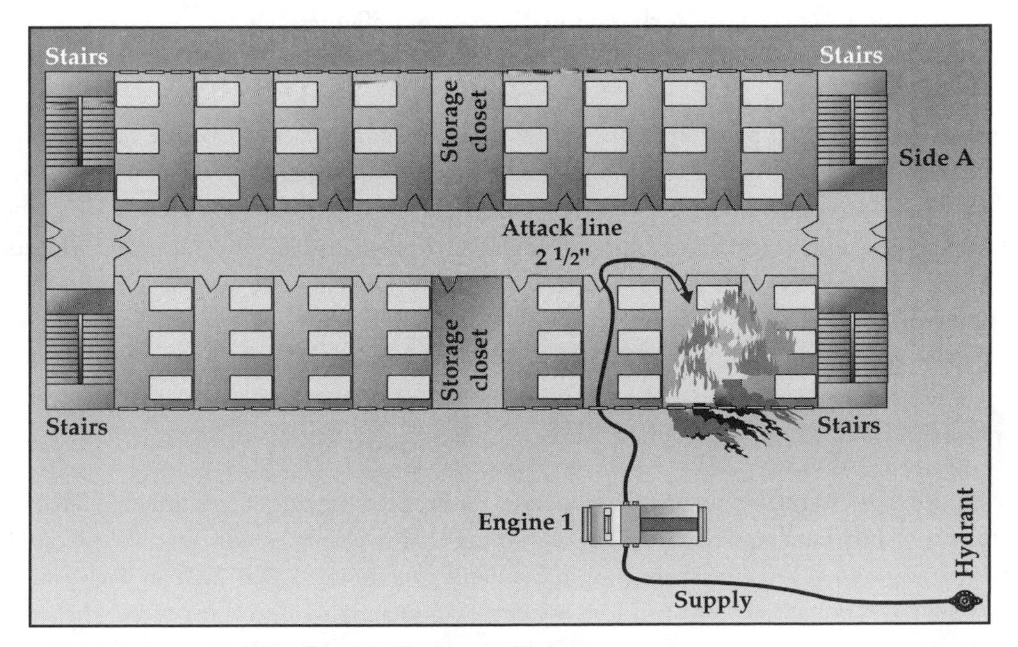

Initial line placement in the Small-Department response.

toward the front of the building. This should ensure the best line placement to protect the most people. Meanwhile, the crew of the second engine lays a 2 1/2-inch line into the rear of the building. This will serve as a backup line to the hose that came in through a window along the side. An aggressive and hopefully adequate attack has now been mounted on the first floor. That's the first leg of a defend-in-place strategy. The next task is to begin ventilation. The IC assigns a crew to open up the front of the home. In a larger department, a truck crew would bc assigned to this task. In a small department, you take whoever's available. A RIT should be established as soon as possible.

Once the fire has been darkened and crews are venting on the first floor, searchers are assigned to search Division 1. As soon as additional members become available, they'll be assigned to search (for life and fire) and overhaul the second floor. Due to the severity of the fire, rescue groups are assigned to facilitate the removal of the victims.

Secondary assignments can follow the removal of victims to a safe haven. First and foremost among these is the triage of all residents. Many of these people are in a fragile state of health. Oxygen will be required in great quantities. Many of them will be confused and apprehensive. If any health-care experts are available, including those specializing in mental health, you should call them in to render assistance. Find another building to stage any and all residents who don't require medical attention. Many of these people will still require close supervision and monitoring.

After all of the areas of the building have been given a secondary search, the investigation as to the cause of the fire can begin. Any crews that aren't still busy can take a break at this point. Set up a fire watch in areas of heavy involvement. Thc line used for the attack can now bc wyed off to 1 1/2- or 1 3/4-inch mop-up lines. Two firefighters will be needed for this task. After the cause has been determined, final extinguishment can begin.

With fires of this magnitude, the chief won't normally be leaving while his crews are still picking up. There will still be many issues to discuss with the management of the nursing home. Included among these is the question of temporary housing for the residents. It isn't uncommon for a very small community to open up its firehouse and line it with cots. These aren't people whom you can leave standing at the curb after the last length of hoseline has been rolled up and loaded. The chief's resource book can be invaluable at times like this. These are the incidents in which a department can truly prove its worth in the eyes of the community.

Medium-size Department

Staffing will still be the major concern. Most medium-size departments don't have enough on-duty staffing to accomplish everything that needs to be done at a worst-case fire in a nursing home. Second alarms should be mandatory at any fire in such an occupancy.

As before, the initial crews must be focused on attacking the fire and venting the smoke. The initial lines must be placed to protect the most victims. If the officer finds that residents

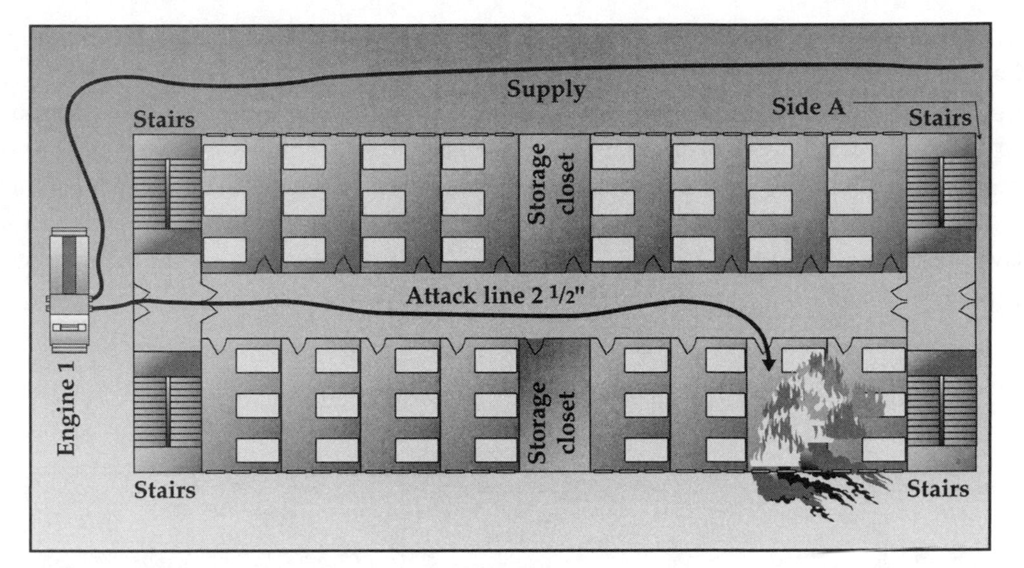

Initial line placement in a Medium-size Department.

are trying to self-evacuate, he should assign available members—from a heavy-rescue squad, for example—to rescue duty. Note that this crew is not assigned to search. The word rescue implies the removal of victims who have already been located. Rescue is a rapidly paced operation. Search takes time. The focus of the rescue crew is to follow the line inward and assist anyone in sight. Until more units arrive or the fire has darkened, the staffing situation will still be too critical to assign members to search.

As second-alarm units do arrive, however, the IC can direct their crews to search areas where savable victims might be. From the description of this fire, this would immediately indicate the rear portion of the front wing on Division 1, the entire rear wing of Division 1, and all of Division 2. Other members will have to check for extension. Finally, many EMS personnel will be needed to triage, treat, and transport the injured. This can tie up an entire second-alarm response or more, depending on the severity of the conditions inside.

Large-Department Response

Because of the number of firefighters responding, concurrent operations can take place. While the attack teams are laying line between the fire and the residents, members of the first-in truck and the rescue squad can start search operations on the first floor. An outside vent man from the first truck, meanwhile, gives the by-products of the knockdown someplace to go. The crew of the third-in engine can take a line to the second floor to check for conditions there, along with members from the second truck.

Reports indicate that there are many victims on the first floor, and the IC can respond with members from the third alarm. He may also need to send another line and hooks to the second floor. Once the fire has been darkened, ventilation commenced, and search conducted, the normal secondary assignments can be addressed. Once the majority of the victims have been triaged, units of the third alarm can begin to return. Treating the victims, overhauling the fireground, and securing the residence become the main points of focus.

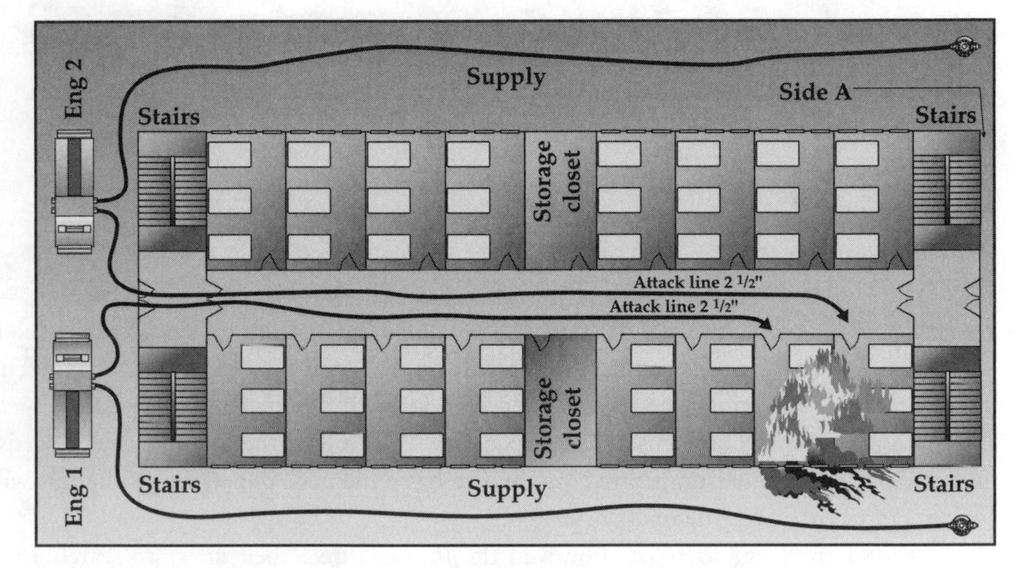

With the Large-Department response, staffing allows for the placement of 2-1/2" lines.

Chapter Sixteen Questions

1. Far and away, the biggest problem associated with nursing home fires is ___________.
2. It would be expected that old, converted nursing homes are fully sprinklered. True or False
3. Given the potential for a large life hazard at nursing home fires, what is the IC's primary concern?
4. The rule of thumb for the number of rescue groups is ___________________.
5. The rule for initial line placement in a nursing home fire is ________________.
6. Backup might be ruled out due to staffing concerns in a nursing home fire. True or False
7. RIT may not initially be possible due to staffing concerns in a nursing home fire. True or False
8. The fundamental reason for a defend-in-place strategy at a nursing home fire is ___________.
9. RIT will generally enter a building ___________________.
10. What is your reason for the answer to question 9?

Questions for discussion

a) Compare and contrast the need for additional staffing to attack and vent a fire in a nursing home vs. the need for RIT and RITs effect on other firefighters.

b) The author discusses corralling and controlling evacuated residents of a nursing home. With only one or two firefighters required for this task vs. additional crews to help search for and remove victims, is this a waste of manpower?

c) How best to protect property and people is dependent upon initial line placement. Defend that statement to those who advocate speed over purpose of the first line.

Chapter Seventeen

Fighting Industrial Fires

Fires in the types of occupancies to be described in this chapter can be some of the most challenging incidents to which we respond. They can quickly evolve into multifaceted emergencies involving both fire and haz-mat concerns. Specific branches to deal with each of these areas may be required under the Operations section. Many times, a third branch, dedicated to evacuation, must also be established.

A great economic impact can accompany these fires. Workers can be sent home for a day, a week, or forever. The tax base of an entire community can be permanently altered. As influential as they might be, these are hardly the bread-and-butter fires of the fire service. For most of us, waging a battle in an industrial setting is something we do only on occasion. As such, we have a hard time picturing what the scene will be like. Crawling around on carpeted floor in a residence is different from crawling on a floor of painted concrete covered with oil. Snuffing out the flames in a couch is different from trying to subdue a river of some exotic flammable liquid.

Some of these occupancies can be huge. The ceiling heights are much greater than they are in residential buildings. Any number of impediments can hinder an emergency response. Even draft curtains hanging from the ceiling can deflect the strongest of hosestreams, rendering them ineffective. The overriding issue at any of these facilities, of course, if the presence of hazardous materials, known and unknown. You may not be able to get proper information about them at the scene. It isn't uncommon for industrial workers to adopt a casual attitude about the chemicals they deal with on a daily basis. Some of those who've been handling the same unpronounceable stuff for twenty or thirty years will assure you that it hasn't hurt them yet, but their bloodshot eyes, crimson skin, and stilted breathing tell otherwise. Don't wait for a doctor or an investigative reporter to refute such statements; don't expect that OSHA or

This three story ordinary building is now used to store automobile tires. *Credit Coleman*

some other regulatory agency has made the worst of these places safe for children and other living things. If something doesn't seem right, feel right, smell right, or taste right, against everything that other people have told you, take caution. Put on your SCBA and pray to God that you'll still be breathing with your own lungs thirty years from now.

The workers, of course, aren't to blame for the horrendous ingredients used in their facility, and they are just as vulnerable to fire as the residents of a nursing home—perhaps even more so when vessels start to explode. In searching an institution or a private residence, firefighters at least have an idea as to where victims might be found. In an industrial setting, the victims might be anywhere, and these buildings never boast a predictable symmetry to make it easier for you to find them.

The construction types of these occupancies vary, of course, but they follow the same essential principles and styles as other commercial structures. You'll find an abundance of ordinary, concrete, and steel construction. These buildings are meant to be functional, not pretty. In newer communities and those with tighter building codes, you'll find some with more outward appeal. Always remember, however, that behind the facade, the same sorts of dangers lurk.

This is only a portion of a new Jeep manufacturing plant being built in Toledo. *Credit Coleman*

An old 4-story ordinary furniture manufacturing building. This now also has two floors dedicated to furniture re-finishing. *Credit Coleman*

Many roof assemblies on manufacturing buildings have sky-lights that will facilitate ventilation. *Credit Coleman*

The haz-mat concern is a matter of scale. These occupancies can all too easily spawn large-scale events. The inherent fire loads and abundance of raw flammables can contribute to large, fast-moving fires. You'll have to be aware of the runoff as never before. The runoff must be checked for pollutants, and the air downwind of the fire will also have to be monitored. Anticipate evacuations. These fires can generate an enormous public response. Smaller communities will almost certainly have to call in mutual-aid forces to help handle the public repercussions of the event. The problem won't be one of simply getting people out of the way. It's deciding whom to evacuate, how to get them out, what to do with them while they're away, and how to protect what they've left behind.

The safety of firefighters and civilians is the first concern of the incident commander. The decision as to whether to mount an offensive or a defensive operation must be made quickly. Interior operations require more staffing in most instances; however, you'll probably have a better chance of controlling the fire and lessening the effects of smoke and runoff. If you can make a quick, aggressive assault on the fire, you may be staving off the nightmare scenario, up to and including public evacuation and ineradicable contamination of the environment. If you can't knock it down quickly, you probably won't be able to knock it down at all—not until it runs out of fuel.

A multiple alarm fire in an occupied manufacturing facility in New Jersey. *Credit Ron Jeffers*

An occupied manufacturing building in Passaic N.J. requires multiple alarms to handle. *Credit Ron Jeffers*

In determining the additional resource needs, go back to the basics and your department's capabilities. How many firefighters are needed to flow 3,000 gpm? How many engines can you muster to maintain such a flow? How many searchers will you need to comb through a 10,000-square-foot area cluttered with machinery, storage bins, and racks in fifteen minutes?

Many of these occupancies have built-in fire protection systems, including automatic sprinklers and standpipes. Many of the standpipe systems are horizontal systems, with hose stations marked on columns in a line throughout the plant. No matter how you plan to get water on the fire, you must still determine how to move in. As always, the most correct launch point is one that will cut off the fire from the rest of the facility and not push its by-products toward other personnel. If you can see the flames, the fire will be easy to fight. The hardest fires to fight are those concealed by dense smoke. If the flames are venting through a window or loading dock, move in on them through the interior and blow them out the portal. If you can't see the flames, only boiling smoke, then take in the line from the area where the smoke appears to be the lightest, either in color or density. Sometimes these signs are obvious and sometimes they're not. If the smoke appears lighter in a particular area, chances are that the heaviest involvement is somewhere else. Victims could stand a better chance of survival in the areas of lighter smoke.

Thermal-imaging technology is a great help in such instances. Look inside with the imager and head toward the regions that show lighter shades of grey. A thermal imager is so useful that, lacking one, you may even wish to hold off all interior operations until a mutual-aid unit can bring such a device to the scene.

If you're not comfortable sending crews inside, ventilation may help. If the building is one story tall, if the fire is on the top floor, or if the fire is in the walls, put a crew up top. If the roof is neither spongy nor sagging, these members can cut a hole with a power saw. The hole should be centered on the area or wing where the smoke is being generated. As the smoke pours out of the hole, the fire may light up, allowing you to take in lines from the most opportune point. If topside ventilation isn't safe, pick a side of the building where the by-products of the fire will cause the least harm, then open a few windows and prepare to come in from the opposite end. The fire will move toward its source of oxygen, and you can push it toward those portals with your hoselines.

Topside ventilation is an oldie but a goodie. It has stood the test of time. With today's power tools, this operation can be performed quickly. Many industrial facilities, of course, have skylights, and these should be your first choice unless they'll drag the fire to an uninvolved area. Sklights are easy to take and they're cheap to replace. Many of these buildings have sawtooth roofs and numerous skylights. Take advantage of your good fortune if the building has a sawtooth roof with a series of vertical windows. Break the entire row of windows as directly over the fire as possible. Once the fire is under control, you can expedite ventilation by closing as many ground-level doors and windows as possible, then turning on a PPV fan.

Monitor roofs and skylights should aid ventilation efforts. *Credit Coleman.*

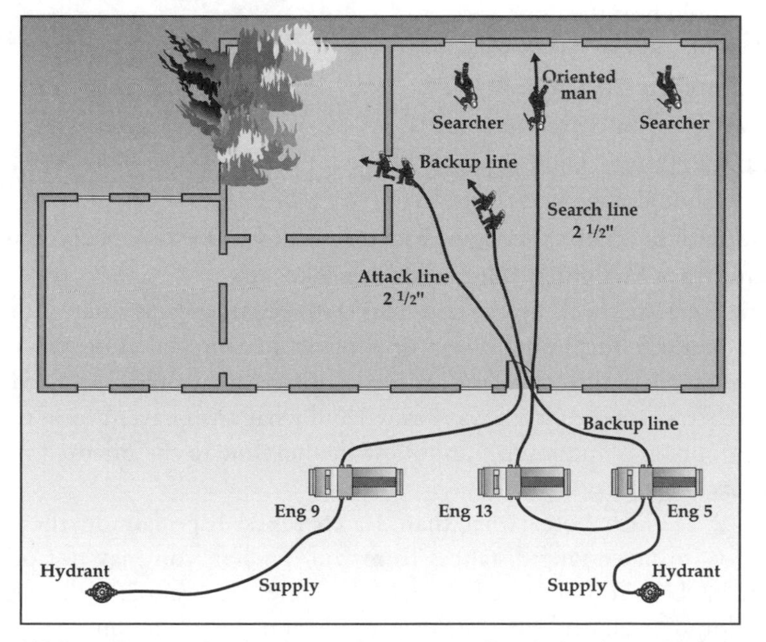

While a line works the fire and a backup line is in place, Search pulls its own line to search off of at a fire in a manufacturing plant.

RIT will note the fire escapes up side "C" of the concrete structure. *Credit Eric Renzhofer*

Positive pressure is the fast and effective means of venting a factory. This is especially true if you've located the fire. The size of the occupancy, of course, is a factor. I once responded to a fire in a large automobile-transmission manufacturing plant in Toledo. We were confronted with a very hot, smoky fire in the ductwork of an oil vapor collector in the center of the facility. We ended up using several PPV fans and two hovercraft on trailers, but shortly after the operation commenced, the smoke was no longer a problem. We cleared out an area of more than 100,000 square feet in about fifteen minutes.

When the visibility is low in these buildings, the oriented method of search works best. Several crews, each using a 2 1/2-inch line stretched to the limits of the interior, can cover a large factory quickly and safely. Connected to an engine or a manifold near the door, each line can reach up to about 250 feet into the building. Searchers working off 30-foot leads of webbing can cover an area measuring 60 X 250, or 15,000 square feet. Each crew should consist of three members, at minimum. Thermal-imaging cameras work well in these operations, too, and are especially useful when firefighters must scour mazelike areas.

It's essential that you back up the searchers with rescue groups in an industrial setting. There would be far too much duplication of effort if the searchers were also responsible for dragging out whatever victims they find, and additional time would be wasted in describing to another searcher where the victim was located and what areas haven't been covered. A two-person rescue group, as a minimum, can follow the hoseline to the oriented man, where the victim can be exchanged.

Please use 2 1/2-inch hose rather than search ropes! Especially in these occupancies, where you'll have to move long distances from your anchor, you may not otherwise know when your line has been compromised. Also, backup lines are essential at these fires. The position of the backup line depends on the crews inside and their function. If only attack crews are present, then the backup lines should follow and protect those members. If searches are in progress, then the backup team should shadow the searchers.

In these surroundings, the backup crew should consist of at least four members, and the nozzle should be positioned a little closer to the front team than I normally recommend for other fires. This will increase the distance between the backup nozzle and the point of entry. This area needs to be monitored by backup members. That's part of their responsibility—to ensure a safe means of egress. Four members are required, since two can occasionally travel back along the line to the entrance, ensuring the viability of the path, while the other two remain at the ready on the nozzle.

Because of the many hazards, the members of the RIT may want to have in their possession any information gathered during the pre-incident stages; specifically, a description or plan of the building, including its hazards and anomalies. Having up-to-date information is essential. Many industrial facilities are continually evolving, and a single addition tacked onto the rear, out of the sight of the road, can make all initial rescue plans obsolete. Depending on what's inside the building, the members of the RIT may need to take a crew with a hoseline with them. If flammables, explosives, or other volatile agents are present, these members may only become part of the problem if they aren't able to defend themselves. If you, as Command, don't feel comfortable sending in more crews, don't send them in. And if they do go in, at least give them some protection.

This cold-storage building is made of concrete construction. This building is similar to one that took the lives of six brothers in the fall of 1999. *Credit Coleman*

One more word needs to be said about the guarantee of safety for personnel. A firefighter who spends two hours in a contaminated atmosphere may suffer an extreme exposure, whereas an evacuee who only spends two minutes in that same environment might suffer no ill effects at all. You, as a firefighter, should be prepared to explain to any evacuee why he may not need protection but you do. A particle mask may do nothing to filter out the effects of a dangerous gas; still, by offering a particle mask to an evacuee, you may remove the question from his mind. In the event that you are asked, you can answer that he will be in that contaminated atmosphere only for a matter of another few seconds, whereas you may be operating there on and off for the next several hours. Such an explanation and an authoritative presence will likely serve to expedite the rescue without your having to surrender, even for a moment, control of your SCBA. Never forget that the air in your bottle is your most precious, most scarce resource.

SCENARIO

Although it has changed hands a number of times, the building has always been a manufacturing facility since it opened in 1901. Its original workers turned out hand tools. In

The simulation for this chapter occurs in the ordinary (CMU and brick) building with a saw-tooth roof that houses a chemical manufacturing plant. *Credit Coleman*

1922, this business closed, and the building opened two years later as a chemical plant. Although it has undergone three takeovers since then, its owners have all been in the chemical business. The building has been extensively renovated and enlarged from its original form. The company that resides there today manufactures rubber latex gloves for hospitals, as well as resins and salves used in the health industry. Polybutyl rubber, phenol, and other chemicals are stored in quantity.

The building itself is of ordinary construction. The exterior load-bearing walls are of CMU and brick. There is a full basement with a floor of poured concrete. The walls of the basement are made of stone, and in some areas around the foundation, sections of the original stone foundation still exist. There are windows along all four sides of the basement. A saw-tooth roof supported by steel trusses tops off the structure. The sloped side of the teeth are sheathed in plywood, tar paper, and rolled asphalt. The vertical faces are made of glass and steel. Because of grandfather clauses in the codes, there are no built-in fire protection or detection systems on the premises.

There is a significant exposure on Side D—a group home; a temporary shelter for women. There are eight individual rooms that can house women, plus up to three children, if necessary. There is a living room and kitchen on the first floor. On the day of the fire, there are twenty-eight women and children in the home.

The fire occurs on a cloudy Sunday morning in the spring. A chemical reaction spawned by an overturned beaker has been slowly brewing over the weekend, and now it bursts into open flames.

Small-Department Response

The first officer on the scene reports thick, black smoke showing from the industrial facility, but the size of the fire cannot be gauged from his vantage point. To be sure, crews will have to get inside. The first assignment is to stretch a line inside and check the extent of the fire. Because of staffing limitations, the first two engines are assigned to work as a single unit. This provides at least four members to advance the attack line, a driver to operate the pumps, and an incident commander.

The initial line goes in through the front door, as long as the fire isn't in that immediate vicinity. For reasons of speed and mobility, the members use 1 3/4-inch line. If the location of the fire were known, a 2 1/2-inch line would probably be the better choice.

To alleviate the smoke conditions inside, the IC assigns his truck members to take out panes from the saw-tooth roof. Warning is given to the members inside so that they won't be hit by falling glass. Once the vent crew opens up a hole over the fire, they leave the roof. This is no place to take a break.

The incident commander doesn't personally perform a walk-around of the building because of the size of it, but he sends the members of an ALS unit to the rear to check on some errant smoke. These members report back that the fire is seated in this vicinity. At the

The fire as indicated in this chapter's case study. This is side "A" and "B" of the building.

Sides "D" and "C" of the fire building.

same time, the interior crews report that they need additional lines inside. The IC responds by issuing a second alarm.

As units arrive, the incident commander assigns some to the interior. If the operation turns defensive, or if anyone in the group home becomes sick from the fumes, he will assign other members to begin evacuations. He also assigns a RIT and if necessary will have that crew shadowed by its own backup team. Once the spread has been stopped and the residents of the group home have been treated, crews will then surround the fire. Heavy equipment will be required for the overhaul. The later units will be used for the mop-up phases; the earlier crews will go through rehab and be sent back to the station. Due to the extent of the fire, the time of day, and the day of the week, no primary or secondary searches were performed. Sometimes you just can't get to everything.

Medium-size and Large-Department Response

With the location and extent of the fire uncertain, the incident commander of a larger department would probably still do best by committing a maneuverable line to the interior. Having more manpower from the outset, a backup line is also in order. Ventilation is accomplished by the same means: breaking vertical panes in the saw-tooth roof. With reports from the interior of an advancing fire, the IC requests a precautionary second alarm. An additional line is sent in to replace the backup crew, as the backup crew joins the attack. A RIT is also established.

A larger department is more likely to have a haz-mat unit, and these members should probably assist in monitoring conditions in the group home. Additional units are sent to triage and treat the ill. As the bells of the first-in SCBAs begin to go off, the IC enjoys the luxury of being able to send in fresh members. He'll need to set up a rotation until the fire can be brought under control. If and when the evacuation crews in the group home complete their assignment, they can also be used for the main operation, since these members are relatively fresh.

Overhaul will be the most important secondary assignment, but here again, a larger department should be able to provide sufficient personnel without having to assign members from mutual-aid companies. Salvage in areas remote from the extreme heat will also be required. Computers and paper records in office areas should be preserved and protected.

Chapter Seventeen Questions

1. Fires in industrial occupancies often involve two fronts. Those are the _____ and ________.
2. Workers at industrial occupancies can provide the best information concerning the effects of the hazardous material on site. True or False
3. Runoff will be a major concern at a fire involving an industrial occupancy.
4. The __________________ is the first concern of the IC at a fire in an industrial occupancy.
5. According to the author, fires where _____________ are the hardest fires to fight.
6. Skylights in industrial occupancies are excellent in venting the fire. True or False
7. The author advocated the use of a _________ over a search rope in a fire in an industrial occupancy.
8. Why is a four member back up crew necessary in these fires?
9. Should the backup crew be a little closer or a little farther away from the attack crew in a fire in an industrial occupancy?
10. Who should check to see that the way out is clear and free of fire?

Questions for discussion

a) The author states that fires where no visible flame is present upon arrival, only heavy smoke venting from the building, are the hardest fires to fight. Please discuss this concept.

b) Evacuations can be extensive operations at these fires. Discuss a civilian wanting your SCBA during an evacuation and how you would avoid this problem.

c) Discuss "focus" as it relates to the fire and the hazardous materials problems at a fire in an industrial occupancy and how IMS will help this split in Command's focus.

Chapter Eighteen

Fighting Fires in Hotels

By code, most of these occupancies must be protected by a complete automatic sprinkler system. In part because of this, working fires in these occupancies are rare. Still, many high-casualty incidents have occurred in these buildings. The text that follows is based on a worst-case scenario, in which the sprinkler system has malfunctioned. Also, in the general discussion, I won't differentiate between a hotel and a motel. The distinction that I will draw is whether the building in question is a low-rise or a high-rise, and I will point out some characteristics related to age.

Life safety is your major concern, but panic may be one of your biggest problems. If the building is equipped with a public address system, use it to inform the guests of the situation. Try to control needless evacuations. Any good front desk should instantly be able to provide you with a list of the occupied units, but the number of occupants in a restaurant, banquet hall, or recreation area may be more open to question. If such facilities are open during the time of the fire, it probably won't be easy to account for the people in them. Your need for EMS resources may be huge. At minimum, the sectors that you establish must include an EMS branch with triage, treatment, and transport groups. Within the transport group, there should also be a transport staging area and a transport staging officer.

In terms of construction, these occupancies can be divided into two basic categories: old and new. For ease of discussion, I will consider anything built before 1975 to be old, since that is the grandfather date for built-in fire protection in these occupancies.

The exterior load-bearing walls in old hotels can be of concrete and steel, and thus truly fire-resistive, or they can range to ordinary and balloon-frame construction. A very old hotel in a nationally recognized amusement park near my home town is of balloon construction. Hundreds of guests go to sleep every night in this century-old hotel. I believe that it has been

A modern low-rise motel. Would code require this building to be protected with automatic sprinklers? *Credit Coleman*

A fire in a high-rise Hotel. What are the initial concerns at this fire?

retrofitted with an automatic sprinkler system, but the same assertion can't be made for others of its type across the country. Old low-rise hotels generally have wood-rafter roofs. Some of these are flat with a cockloft, and some have gables. Very old low-rise gable roofs have either plywood or purlins with asphalt or wooden shingles. Flat-roof assemblies on these very old low-rise hotels have plywood or plank sheathing with tar paper, tar and gravel, or a newer membrane covering. You'll find mostly wood joists in the floor assemblies, with hardwood, plywood, and some plank floors. Some old one-story concrete motels sit atop a concrete slab.

New low-rise hotels are of wooden platform or truss construction. One-, two-, and three-story versions generally have a center court or wing layout. High-rise types have steel, concrete, or combination exterior load-bearing walls. To my knowledge, there has never been a collapse of a modern high-rise hotel due to fire in the United States. These buildings are made to withstand the effects of open flames, although, as we all know, severe interior damage can still be the result. New low-rises with flat roofs have either tar and gravel or membrane roofs supported by trusses. Gabled versions have truss assemblies with asphalt shingles or terra cotta tile. Some have resistive shake shingles. Newer low-rises will normally be on a concrete slab, but those with basements or lower levels tend to have wood-floor assemblies with either truss or engineered I-beam construction. Usually you'll find plywood sheathing and carpet over the trusses. Modern high-rise hotels generally have concrete or steel roof assemblies, with built-up metal-deck or membrane roofs, and the floors, if not all concrete, have poured concrete over Q decking.

An old three-story Hotel. This building lacks any built-in protection. *Credit Coleman*

Fires in low-rise hotels generally spread in a fashion similar to that of garden apartments and other modern wood-truss buildings. With decent drywall and taping, these buildings can resist flames for quite a while. The windows and doors tend to be the weakest links. Fire can spread vertically in plumbing voids and pipe chases, especially if the workmanship is shoddy. Remember that plumbing chases will generally be in a vertical line up to the top floor, then out through the roof in the form of a soil stack. You must check this avenue early when confronted with a working room-and-contents fire. Fire can also spread vertically by lapping out of a window and heating the room above through the window until materials inside reach ignition temperature. This process, also known as autoexposure, can be a significant means of extension in these types of buildings. Fire will also take advantage of doors left open by fleeing guests. Although the floor coverings in these occupancies are rated, they'll still contribute to the problem if heated enough.

A fire in a high-rise hotel won't generally threaten the entire guest population. Guests three or more floors above the fire will likely be safer in their rooms than they would be trying to flee by way of the fire tower or stairs. It's imperative that you get two firefighters on each floor as soon as possible, both to control the activities of the guests on the upper floors and to monitor conditions. The principal avenues for smoke travel in a hotel are the hallways, stairs, elevator shafts, and HVAC systems. Smoke will also be forced into imperfect plumbing chases and other such voids. In terms of behavior, you may discover that smoke can skip several floors. A fire on the fifth floor may have little effect on the floors above until you get to the tenth floor or so, which turns out to be dense with smoke. Building engineers on the scene should know the capabilities of the HVAC system and what corrective actions to take in these situations.

For most departments, even the best ones, staffing will prohibit simultaneous tactical evolutions from being conducted. Normally, the initial crews will work off of standpipe systems from stairwells and fire towers. There may be some delay in getting to the floor of the fire to hook up, and then in advancing the line to the fire room. Unless your department can drill on these evolutions frequently, your reflex time will be lengthened simply out of your lack of experience. It's like riding a bicycle once every five years or so. You'll be able to do it, but you'll be a bit wobbly at first and unsure of yourself.

Take a look at the manpower requirements. The average floor space of a typical mid-sized hotel measures 12,000 square feet or more per floor. A typical room measures 300 to 450 square feet. Two rooms can require the flow of one 2 1/2-inch line for good extinguishment. Two 1 3/4-inch lines may do the trick, but you'll need to calculate the pressures perfectly, and some standpipe systems won't support automatic nozzles requiring 100 psi. It'll take a good team of two to handle the 2 1/2-inch line, but it'll take four to manage a pair of 1 3/4-inch lines.

In the best situation, two members will be needed for ventilation, using horizontal ventilation, or perhaps a stairway and PPV. In most cases, four or more will be required, and such a number doesn't yet begin to involve the roof. With a couple of basic search crews, a two-man rescue group, and some backup firefighters, you'll be using about sixteen members at a

This is an example of a "100-man fire". A fire in older Hotels presents many challenges.

This low-rise motel is in essence a wood-frame building similar to platform construction with the deck assemblies being truss. *Credit Coleman*

Where would the initial attack lines be taken in this two-story motel fire?

An old Hotel which has load-bearing structural members predominately of concrete. *Credit Coleman*

bare-bones minimum, and you can't handle a major working fire with that. Obviously, multiple alarms or mutual aid will be required. Deputy Chief Jim Murtagh of FDNY calls these "hundred-man fires." He states that the number one hundred accurately reflects the minimum staffing required if you are to anticipate an optimal outcome. A fire flow of more than 2,100 gpm could be required if 25 percent of a floor in a large hotel were involved. To provide that sort of gallonage would require ten 2 1/2-inch lines, meaning a minimum of twenty firefighters. If they can't place that flow to good effect, their fifteen-minute air supply will expire before the fire does. As they go back down to change bottles, chances are that the fire will reacquire all of the real estate that it held before the attack. In theory, to avoid losing ground, a second crew will be needed to maintain the attack when the air bottles of the initial team run low. This translates into forty firefighters dedicated to the attack alone. To successfully search for sixty or so guests on a floor in fifteen minutes will easily require an additional thirty to sixty firefighters, not counting those already assigned to controlling panicky guests on the floors above. There's your hundred-man response right there.

In a low-rise hotel or motel, your initial attack will likely be similar to the traditional attack used at most structure fires. Expect to pull preconnects for speed. Most motels have a parking lot that surrounds the building, and stairways and interior doors are placed at intervals for the convenience of the guests. Firefighters usually gain access to the second and third floors by means of the stairs, whether inside or out. Most 200- or 250-foot preconnects will reach the remote areas of these buildings, though pre-incident planning will give you a firm answer to this. Some of these occupancies also have standpipes. Take the shortest approach that will allow you to cut off the fire from the majority of the guests. A ground ladder to the balcony of an unaffected room several doors down may be the fastest way to get to the fire room.

Think strength. Always take in more than you'll need. If you think that a single 1 1/2- or 1 3/4-inch line will do, take in a 2 1/2-inch line. If you think that one 2 1/2-inch will do, take two. Position the initial line with a simple objective: to stop the spread and confine the fire to the smallest area possible. Secondary lines can be taken in from tower ladders. Secondary lines to check for extension can be taken in from standpipes. Secondary lines for overhaul can be taken in through the window of the fire room. Don't experiment with a hotel fire as a drill tower training ground. Stick to the basics and put out the fire.

Attacking fire in a high-rise hotel is another story. Strength and simplicity of deployment will still apply, but speed won't be a positive factor. In a high-rise, a standpipe is the tool of choice. From your pre-incident planning, you should already be familiar with their location, their connections, and whether you must contend with any pressure-reducing devices. Also, you must know the rating of the system. Many of these are designed to provide only 65 pounds of pressure at the upper floors. Many automatic nozzles are designed to operate at 100 psi. If you're not prepared for this discrepancy, you may be in for a shock when you open the bail of the nozzle.

Typically, an attack crew takes in a line and knocks down the fire in a few seconds, but what if you don't? What if you advance on the fire but your advance is slow and it seems you're just pushing the fire ahead of you? Now the alarm on your SCBA goes off. You correctly decide that it's time to back out to the stairway to change bottles. If you back off, however,

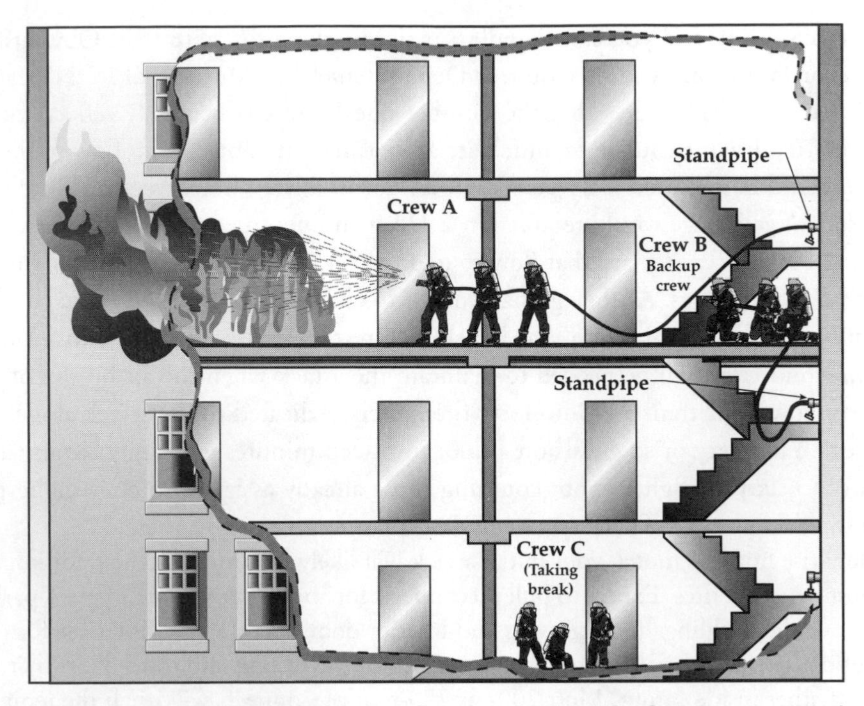

A three crew attack evolution in a fire in a new motel. Crews are rotated between attack, back up and rehab.

the flames may eat their way right back to their former position. To prevent this from happening, you need to plan ahead with respect to manpower so as to mount a sustained attack. You can do this by assigning several crews to one line. As Crew A works the fire, Crew B stays in the stairwell with their SCBA in a standby position. If Crew A gets into trouble, Crew B can follow the line and provide backup. If the members of Crew A need to change their bottles, Crew B can move up the line with fresh bottles and take their place. Crew C, meanwhile, two floors below, can stand by at the ready with SCBA and tools. When Crew A comes out, these members can take a breather two floors below, and Crew C can move up into the stairwell. Ten or twelve minutes later, they all shift positions again. A three-crew rotation is the only way that I know of to sustain an effective attack in a high-rise.

The first rule of search is only to attempt it if you have the manpower to do so. The second rule is to concentrate on those who aren't making any noise. For the most part, people yelling at you from balconies with light smoke wafting through their open sliding doors aren't in any real danger at the moment. Talk to them and reassure them, then go look for those who are really in trouble. The third rule of search, of course, is to start in the area closest to the fire where savable victims might be found, then work away from the fire. The cold, hard

Where would you expect the search crew to start their search in this fire?

fact is that we might not be able to save everyone. Let experience and your senses be your guide. If you're getting into an area that's real hot, chances are there won't be anyone still alive there. Don't waste your time on the dead. Move to an area where it's not so hot.

Successfully searching these occupancies isn't an accident. The basic requirements are plenty of staffing, a plan, coordination, and the cooperation of the rescue group. Because of the regularity of the rooms off the hallways, I believe hotels and motels provide the optimum setting for the oriented method of search. One oriented man can control three firefighters, and he knows what's been searched without resorting to crayons, chairs at the doors, or any other marking system.

As ever, time is the enemy, so several crews will be needed at a working fire in a hotel. Personnel should be sent above the fire as soon as possible to report on conditions there. Given a heavy smoke condition, occupants should probably be removed from at least the two floors above the fire. Hopefully you won't have to search for these people. Use the PA system or some other means to alert those above the fire to stay in their rooms. Tell them that firefighters will systematically knock on doors and escort them to safety. Such a procedure should allow for an orderly evacuation of those in danger above the fire.

Establish a rescue group at these incidents. Each team should consist of at least two firefighters and be assigned to a search officer. If several search groups are established, it's expected that significant numbers of victims will be found; therefore, each should have a rescue group.

A two-story low-rise motel. The exterior balcony provides access to all guest rooms. *Credit Coleman*

If a Rescue Group were established at this fire, I would expect it to stage at the stairway door below the fire.

Position each rescue group on the stairwell of the floor on which its respective search team is operating. If there are four members in a given rescue group, they can operate as two teams of two. When the search officer informs the rescue group that a victim has been located, the rescue group should crawl down the hall to the officer, who is the oriented man. As the victim changes hands, the officer should tell the members of the rescue group the number of the room in which the victim was located.

Set up a predetermined triage area two or three floors below the fire. Any information gained from the search team should be given to the triage officer so as to help identify the victim. After the exchange, the rescue group should return to its staging area on the stairwell of the floor being searched. This is tiring work, and the rescue crew should be relieved as often as possible. Remember that they will also be on bottled air, climbing and crawling as they go about their task, sometimes with the added weight of another human being.

Ventilation will best be achieved by using the HVAC system, if it can serve you at all. Some HVAC systems exhaust air to the outside by way of very short duct runs. This may actually aid the ventilation efforts. The next best alternative is to use a fire tower with access to the roof. The tower should be pressurized, either with its own built-in mechanical system or by using PPV fans at ground level. Keep the door at the top of the tower open, and don't use the tower as a route for evacuations or firefighting efforts. Also, open the door before turning on the fan. After the vent hole in the tower at the roof has been opened, you can turn on the fan and open the door to the fire floor. Finally, open the window in the fire area. The smoke will be drawn down the hall by the venturi effect, then up and out the top of the tower.

Venting Low Rise

We have successfully used PPV fans on the fire floor to aid in the movement of air in the hall on the fire floor. We enter an uninvolved room, preferably one at the end of the hall, and place a PPV fan in the front window, pulling air inward. This will also help move smoke down the hall and out. You can move this operation from floor to floor to clear away residual smoke throughout the building.

Backup crews take on a difficult role in hotels and motels. In low-rise versions, firefighters usually gain access via the stairs, and the lines will generally be preconnects. If the doors to the rooms face outward, toward a wraparound balcony, then the backup line will probably come off an engine in the parking lot, and it should shadow the attack team; otherwise, a search team (if one has been deployed) should be shadowed. At this sort of fire, a single backup line should suffice. If the doors face inward, toward an interior hallway, then the backup line can come off of either an engine or a standpipe. In any scenario, the backup team should leave room for the other crews to work. Backup members have no other function than to provide a safe means of egress. They do not provide additional firepower.

If the backup team is going to work in an interior hallway, then the ideal nozzle is one that will provide a fog pattern that can effectively hold the fire in check. A nozzle flowing 200 gpm can keep a lot of fire from advancing down the hall. A smooth-bore will have to be played off the ceiling, and it won't be as effective as a fog nozzle. Outside, either nozzle will suffice. A breakaway nozzle with smooth bore and a fog tip is the ultimate choice. The officer can keep the fog tip in his pocket. If the smooth-bore nozzle can't hold the fire, then the fog tip can quickly be substituted.

A rapid intervention team will surely be needed at one of these fires. It's a tough decision to make, but you'll have to draw a line somewhere on search and rescue, assigning no more members to these operations until a RIT can be put into place. A good commander will tell himself that a RIT is necessary and not a waste of four members when there are so many guests to be rescued. Trust me, your searchers will extend themselves farther and perform more effectively if they know that someone is there, ready to respond for them. The staging location for the RIT will vary. In a low-rise, these members can stage at the command post or closer to the action if the command post is remote from the main arena. In a high-rise, the RIT should stage two floors below the fire.

Can the RIT take on a dual role at these labor-intensive fires? I will never say never; still, these members will need time to gather their tools and conduct a scene survey. Also, they should spend some time developing a plan of action. If they're redirected to act in another capacity—as a rescue group, for example—two problems will arise. First, their focus will shift, and second, they'll already be fatigued when the Mayday goes out. Never dismiss the option, but think about it very carefully first.

SCENARIO

The scene of the fire is a new three-story motel that boasts large rooms with a kitchenette, living room, and a separate sleeping area. Some of the rooms even have a fireplace. There is a bar and small restaurant on the first floor, as well as an indoor swimming pool and a gym.

The complex is actually three buildings connected by a common hallway. The lobby, restaurant, and pool building are in the front of the complex. There are no guest rooms in this one-story section. A single hallway leads off the rear of the office to the longest wing of the building. This rear wing, which runs along Side C of the complex, has fourteen guest rooms per floor. At the B end of this wing is a stairway with a standpipe connection on each floor. On the D end of the wing is a hallway that leads to the third building, which has twelve units per floor. At each end of this D Wing is a stairway, and both have standpipes.

There are two master suites per floor. These are located on the C Wing of the complex on either side of the hallway leading from the office building. There is an elevator between the two master suites. These suites have two bedrooms and two full baths.

The entire motel is built on a slab. All of the exterior load-bearing walls are made of wood, with platform construction throughout. The floors are supported by wooden trusses.

The case study for this chapter is in this three-story wood-frame low-rise motel. The guest rooms in this occupancy are larger suites. *Credit Coleman*

There is plywood and carpeting over the plywood. The roof is supported by triangular trusses, and it is covered with plywood and asphalt shingles. The roofs on the C and D Wings are hip roofs. There is one access hole to the trussloft next to the elevator on the C Wing.

The building is fully sprinklered. There is a connection in the rear of the building on D Wing next to the rear exit. There have been problems with this system. It was installed in a hurry to meet the demand for rentable rooms. After the motel opened, maintenance personnel began noticing leaks at many joints along the system. The system is shut down during the day so that workers can try to fix the leaks. A fire watch is posted during these hours. This watch is done by off-duty firefighters. With the fire watch in place, the motel has been allowed to remain open for business.

On the day of the fire, the motel is filled to capacity. The department receives a call from the off-duty firefighter assigned to fire watch. He alerts the officer of Engine 1 that he thinks he smells smoke in D Wing. "It comes and goes," the young firefighter tells the lieutenant. It's Friday morning. Plumbers have been working on the sprinklers since 5:30 a.m. Soon after the report from the firefighter, the desk clerk receives a call from a guest on the second floor who reports that she can see smoke from across the hall. The fire watchman heads up the stairs to the second floor in D Wing and encounters thick, black smoke. He closes the door and calls the dispatcher via portable radio.

Small-Department Response

This is a worst-case scenario, and staffing is clearly the critical issue. It's essential that you do the most with what you have. The commander could order an offensive attack. Using all of the on-scenc crews, he might be able to get two lines into play. However, lacking adequate ventilation and a place for the fire to go, this strategy may fail. Additionally, with only a couple of engines and a truck or two, there won't be enough manpower to mount search-and-rescue operations. This would seem to indicate a defend-in-place strategy, but if the steam from the knockdown isn't given a place to go, then the conditions for the civilians inside could be made even worse. It's for this reason that the commander opts to reduce the attack power and instead decides to vent the building properly. The truck crew is told to open up the second floor and to do so with the stick. Meanwhile, lines will be taken in to protect the greatest number of civilians.

As more units arrive, additional lines are stretched to confine the fire to the second and third floor at the A end of the D Wing of the motel. Rescue groups can be established to

The fire in the case study as seen from sides "A" and "D".

remove those victims located by suppression personnel and other members inside. Finally, this sort of construction lends itself to extension; walls and ceilings need to be opened up as soon as possible to ensure that fire isn't running through hidden channels. If conditions warrant interior operations and the conditions are such that crews must operate on multiple floors or over large areas, the IC must consider deploying a backup line. With low staffing, two firefighters can handle this line. As always, a RIT should be assigned as soon as staffing permits.

Rotation for tired crews will mark the beginning of the secondary assignments. At this fire, fourth-alarm units are used to replace crews who need to change bottles and get rehydrated. Establish fire watches early. If you can't assign two firefighters, then consider a single member with a portable radio and at least twenty feet of visibility to be a minimum. Secondary searches, along with some system to help account for the guests, will be required. Hotel records will prove valuable in this regard. Start the secondary searches in the area of most involvement and move outward. Take care not to disturb any evidence in the area of origin. If investigators are on the scene, they should collect evidence and samples as soon as possible so searches and overhaul can begin. Overhaul will be required, and you should establish fire watches for a considerable time after the event. Rekindles occur frequently in these circumstances, as well as this type of construction. If any fatalities occur, the coroner and clergy will be on the scene. Finally, you'll need to contact other motels in the area to house those who have been displaced.

The fire in the case study as seen from sides "C" and "D".

Medium-size and Large-Department Response

The first officer to arrive at this fire has several things running through his mind. The first is staffing, and he can deal with this by immediately asking for a second alarm. His next concern is for the safety of the occupants of the hotel, and the last big concern is for a quick knockdown. Hopefully, both of these latter issues will be addressed by an aggressive attack.

The commander orders an attack line to Division 2, assisted by the second-in engine. The truck will provide the fire with a place to go. These members set up where Sides A and D intersect on D Wing, and they open the windows along both A and D. They then set up PPV from the opposite side of D Wing.

The squad checks above the fire, but no searches are conducted at this time. With staffing constraints as they are, it's best to concentrate on putting out the fire.

Crews from the second alarm assist with relief and start quick ventilation efforts on D Wing. The PPV fans are taken to the rear of the wing. This should alleviate any spread in unwanted directions. The IC doesn't really care if he pushes the fire into some unwanted areas. He has lines in place on the second and third floors. Crews will aggressively open up all concealed spaces as soon as possible. The aggressive PPV tactic is aimed at keeping as much smoke as possible away from any potential victims.

Due to their construction, these occupancies require a lot of overhaul. They're constructed of little pieces of lumber, which are susceptible to early ignition. Overhaul will take a backseat, but only to secondary searches. These secondary searches must be methodical. This isn't an occasion to be missing rooms or even entire wings. If ventilation efforts are effective, SCBA won't be required, and visual searches will be the norm. The hotel staff can help you save a lot of time in locating and accounting for the guests.

In conducting overhaul, pipe chases and ceilings on the upper floors must be opened, and they should be revisited often. It's best to assign at least one crew, maybe more, to rove about the structure, checking and rechecking for fire. The conditions in the truss loft should be continually monitored. This is no incident to stop looking for rekindles. I know of several departments that have been caught rolling hose, only to notice heavy smoke coming from a reinvolved truss loft. In my department, it isn't uncommon for crews to return every hour on the hour to check on the status of these areas.

Chapter Eighteen Questions

1. At a fire in a hotel, __________ is your major concern but ______ may be your biggest problem.
2. Who is in the best position to provide the IC with info on the number and location of guests in a motel?
3. New low-rise motels are generally ____________ construction.
4. Lapping and autoexposure are the same thing. True or False
5. Smoke is pulled from rooms and hallways in a motel up the fire tower by the ____________ effect.
6. The author suggests _________ per attack line in an advanced fire in a high-rise hotel fire.
7. The reason for the answer in question 6 is _________________.
8. When conducting the oriented search in a motel, the point of orientation is the ________.
9. Rescue groups are a must in an advanced fire in a motel. True or False
10. The best location for a rescue group in a motel fire is in the __________.

Questions for discussion

a) Defend the 3 crew per attack line in an advanced fire in a motel/hotel.
b) Discuss the nozzles that your department currently uses and if there are stand-pipes in your community that operate at 65psi.
c) Discuss preplanning and the use of the HVAC.

Chapter Nineteen

Fighting Fires in Professional Buildings

The modern, low-rise office buildings that I will deal with in this chapter are going up all over the country, and they all have basic similarities. Usually they're no greater than two or three stories in height, and some are winding, wing-type, single-story structures. Most of them are occupied by white-collar businesses. Rarely do they have built-in fire protection. Fire codes normally require built-in protection for business occupancies of more than two stories. Given this loophole, most of the one- and two-story versions are devoid of sprinklers.

These buildings seem to spring up in new developments. Some are on winding roads in corporate ghettos; others seem to spring up out of nowhere along our main thoroughfares. Often the businesses within them are clustered according to type, as in a professional building that houses a variety of medical practices. The layout of these structures usually includes a lobby, central hallways, and office suites around the perimeter. Other times, the entire structure is devoted to a single tenant, with the front doors leading to a reception area, then some unpredictable interior maze of modular cubicles and private offices.

Extremely high fire loads and the lack of fire protection top the list of problems in these occupancies, and the life hazard is at its peak during working hours. Early detection usually allows for successful evacuations prior to the arrival of fire companies. Although the life hazard may be greatest during the daytime, many of these buildings are cleaned at night, so fires that occur after hours still require searches. Insomniac workers may also be inside. Unless the parking lot is empty, search operations should be on the IC's to-do list.

The method of construction also adds its share of worries. In terms of construction, there are three major categories of new office buildings: wood truss, concrete, and steel. As we have seen with other types of structures, each will react differently to fire. In many of these occupancies, the exterior load-bearing walls are made of wood. In such cases, expect two-story

A modern office building. This two-story brick veneer building is a platform wood frame building. It is occupied as doctors' offices. *Credit Coleman*

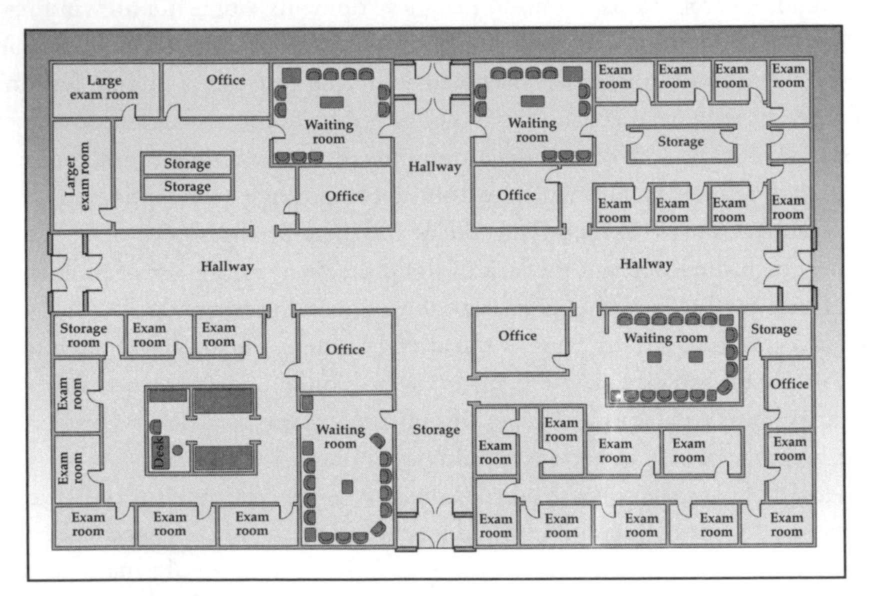

Floor plan for a typical modern doctors' office building.

platform construction with either trusses or wooden I-beam construction underneath. Both of these are bad, especially if the floor spans a wide area below. The exterior finish or siding can be made from a number of materials, including brick veneer, aluminum siding, and shake siding. These buildings may be cavernous within and have a system of columns; otherwise, the structures hovering over the large span may be supported by nothing more than truss assemblies. These buildings, with wide interior spaces and a lack of interior walls, may turn out to be prime firefighter killers in the decades to come. Unprotected truss assemblies that span large openings can surprise crews with sudden collapse. Interior wall assemblies are good! They provide a means of resistance! Roof collapses can be stopped by a system of interior partition walls, whether framed with wood or steel studs. It doesn't matter whether it's a bearing or a nonbearing wall—it will help to stop a collapse. If the engineers and interior designers have been directed by the owners to create an office building with a spacious interior, the roof will be subject to sudden failure. If you aren't sure as to the method of construction and find yourself entering an expansive area, you should inform the incident commander and expect collapse if fire enters the regions above. If the rest of the building is framed in wood, you can expect the roof to be supported by wood trusses. Some of these buildings have triangular trusses to create gabled roofs, and a variety of other configurations exist. The trusses will be carrying either plywood or a cheaper chipboard sheathing, plus asphalt shingles as a top layer.

You'll find many modern office buildings made of tilt-up concrete. These load-bearing walls are precast and then lifted by crane into place. The sections are tied together with rebar, which in some places can be exposed to heat. The rebar will begin to fail at around 800°F. When collapse occurs, whole sections of wall will be pushed outward. Many of the roof assemblies on these buildings are held in place by gravity alone, and the interior partitions usually consist of steel-stud walls, which generally don't contribute to structural stability. Tilt-up concrete buildings normally have concrete roof assemblies, and the first floor can be a concrete slab. A newer cast-in-place concrete floor system is being used in my area. Concrete is poured onto thin steel panels similar to Q decking but with smaller grooves. In the past, plywood forms were used, and then the plywood was removed after the concrete cured. In this new process, the metal decking is left in place after the pour. Regardless, these floors are supported by steel-bar joists and will collapse with 800°F temperatures.

The third type of exterior load-bearing wall, which you'll encounter in the majority of newer office buildings, is made of steel. These walls are normally unprotected steel skeletons supporting an exterior glass curtain or panel walls. The interior walls can again be of wood, but most of them contain steel studs. Buildings of this type usually have steel-bar joists supporting the roof, with Q decking as sheathing and either a membrane or built-up roof on top.

No matter what the exact style of construction, professional buildings are usually furnished and decorated lavishly. These furnishings can contribute to the rapid spread of fire. New plastics can produce tremendous amounts of smoke. Although these buildings tend to be attractive, for the most part, they're built cheaply, and they certainly aren't built with the firefighter in mind. Architects don't spend their day thinking about fire.

This one-story modern office building has a lumberyard in the truss loft. If fire gets anywhere near the truss loft, expect a very intense fire. *Credit Coleman*

Still, high-casualty fires are a rarity in these environs. Although they lack sprinklers, many of these modern buildings have smoke detectors installed during the final stages of construction. Some of these detectors are tied to central-station alarm companies. Early detection isn't a major problem, especially during business hours. Once started, though, the fire will be hot, and it will move quickly along plastic furniture and flammable walls. If it seems as if an undue amount of water is required to darken down the flames, then plastics are no doubt feeding the fire. Some plastics can produce in excess of 12,000 btu's per pound.

The floor plans may be extensively subdivided. Narrow aisles and endless rows of cubicles can be present in every quadrant of the building. Due to the modular design of the office areas, fires in these buildings often aren't confined to small rooms. All too easily, they can roll up to the ceiling and travel horizontally in all directions to the extent of the interior.

If the fire occurs in an abortion clinic or any other occupancy in which family planning is part of the business, you should suspect the possibility of arson and secondary devices. Many of these occupancies don't have signs in front of them, and you'll only know the difference through pre-incident planning. When responding to this sort of occupancy, park the apparatus away from dumpsters and any vehicle in which the engine is running. Beware any packages left unattended at the door. If the incident occurs during business hours, you must consider conducting search operations, and you must be ready to accommodate the police in their investigation of any fire in an abortion clinic.

At the vast majority of fires in a professional office building, attack should be the first assignment, especially if staffing is low. Time given to other tasks may allow the flames to grow and overwhelm the capabilities of most departments. Some of these occupancies can have well in excess of 10,000 square feet per floor. Flows of greater than 800 gpm may be required if even one-fourth of the floor is involved. I would order no line smaller than 2 1/2-inches for the initial attack lines. Forget about maneuverability. Pull in a big line, position it where you can hit the greatest body of flames using an indirect attack off the ceiling, then clean up the hot spots with smaller lines afterward. It would take two additional lines plus twice the staffing to achieve this flow from 1 3/4-inch lines. I realize that you can get more than 150 gpm from a 1 3/4-inch line, but most departments, especially those that use combination or automatic nozzles, don't pump at high enough pressures to achieve it. Most nozzlemen are already complaining of "too much pressure" at a level below that where the good flows kick in.

You should take in lines to hold the fire to its original area of involvement. Correct ventilation is a must in this sort of attack unless you want to be driven out of the building. Once in place, an indirect attack, using a straight or smooth stream off the ceiling, will provide the best results, particularly in a modular environment.

Like high-rises, most of these buildings have suspended ceilings, and there's a lot of stuff in the plenum area—a lot of stuff that can burn intensely. These areas need to be checked often. In conditions of light smoke and fire, you can remove the panels carefully with a hook or some other tool. In a working fire, you can use a hose stream to blast away portions of the ceiling. Always remember that the plenum can convey fire not only above the common areas, but also into individual offices and rooms.

One concern at these fires is a phenomenon that can also occur in high-rise office buildings. This phenomenon is known as the doughnut effect. It can occur in modular office buildings, especially if there is a center-core configuration, where the storage areas and rest rooms are located at the center of the building, and the offices are ringed around the perimeter. In such a case, fire and heat can enter the ceiling, which has few obstructions to slow it down. Hose streams, especially those that produce slight fog patterns, can push fire back and against the exterior walls. If there is no other place for the heat to go, the fire will wrap around the exterior walls of the building, possibly taking hold of the entire perimeter.

Professional buildings that have a number of separate tenants can mean fairly easy search operations, since they resemble hotels, with units at intervals along a center hallway. Use the oriented method of search, follow the walls, and exit at the other end of the building. Given a more complex design, however, search can be far more challenging. Think of your own doctor's office. Mine has a single large waiting room and a small reception area. Behind that reception area is an office space lined with file cabinets. The corridor leads to several exami-

This modern office building has steel studs on the first floor with a lumberyard of trusses in the truss loft. Watch these buildings go up in your community and then fight them accordingly. *Credit Coleman*

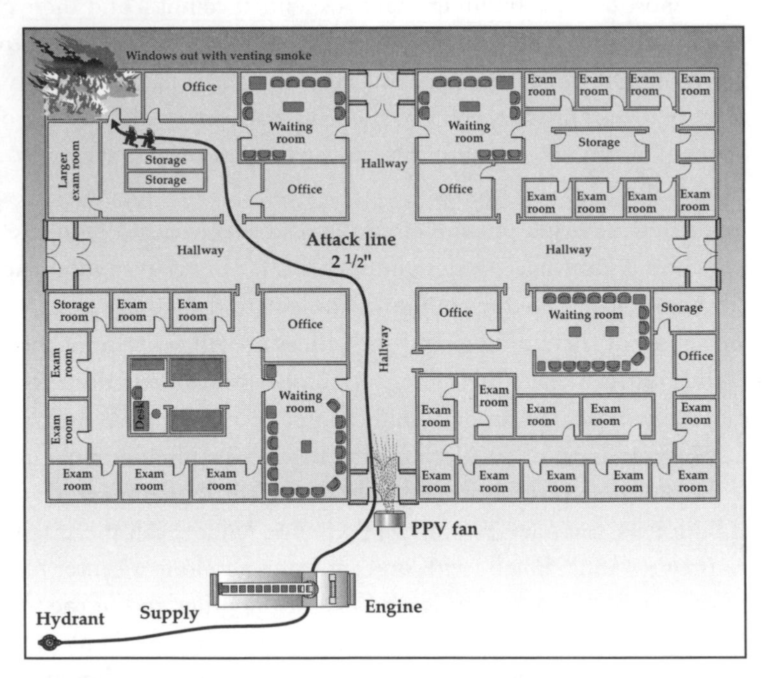

After the fire is located and knocked down, PPV can be an effective tool in a fire in a modern office building.

nation rooms. In my doctor's office, this corridor eventually connects with the the main hallway of the building—it doesn't go back to the reception area at all. How would you search this kind of area in fifteen minutes? How many members would it take to search all of the other units on that floor? You'll either have to do the math and get help, or commit these members to attack and save the searches for later.

Ventilation in these buildings may be difficult or easy. Wood-frame structures with interior hallways and many different occupants will normally call for your department's ordinary ventilation practice. A fire in these buildings is the same as one in a two-story motel with an interior hallway. If PPV is your first tool of choice, then blow the smoke away. Some of these buildings can be quite large, so you may need to stack several fans or take them to specific areas that must be cleared. If you prefer negative pressure, you should knock down the fire, perform a bit of natural horizontal ventilation by taking out a few windows in the rear, then pull the rest of the smoke away.

Topside ventilation must be done with care. If fire enters the truss area, I would use one of two tactics. If there are plenty of interior wall assemblies, pull the ceilings from below and attack from below, if this can be done safely. Let expanding steam extinguish the fire and then wait. In these situations, patience is a virtue. If there are few or no interior walls, then you should let the roof vent itself. Remember that there are two types of building owners: those who have and those who don't have insurance. Neither of these conditions were up to you, nor is it your fault that these buildings weren't designed to withstand the effects of fire. Some of these structures we just can't save. Salvage what you can prior to collapse, then pull out and let it come down.

Some of these buildings have a glass-covered atrium, or perhaps several skylights in the roof. These can be used for ventilation. Safety being a critical issue in this operation, an articulating boom aerial would be an excellent tool to use. It just isn't good to put anyone above or below a trussed area that's already involved in fire.

Steel buildings with curtain walls can be vented by PPV or more conventional means. Again, don't be fooled into believing that steel buildings hold up well during fires. If the steel-bar joists are subjected to temperatures of 800°F or so, you can anticipate collapse anywhere from one minute ago to five minutes from now.

Concrete buildings generally prohibit topside ventilation, since the mass of the roof makes cutting difficult if not impossible. Horizontal ventilation, either positive or negative, will likely be the method to use. In any type of building, if you intend to use the HVAC system, you must know whether it's a closed or an open system. Closed systems circulate air inside and will probably do more harm than good. Open systems take in air from the outside, then condition it and pump it through the building, then vent it to the outside again. These systems can prove helpful in a fire. You should have already addressed this issue during pre-incident planning so that you won't have to wait for the advice of a building engineer during an emergency.

Backup lines are essential when conducting interior operations, and the parameters are the same as for other circumstances discussed in this text. If a backup line has been deployed for the sake of the search team, then it should follow the search team. If the backup line is

A curtain wall exterior hides the interior of this modern office building. *Credit Coleman*

meant to protect the suppression team, then the two-thirds rule should apply. If the building is sizable or if conditions warrant, the nozzle should be positioned closer to the battle lines. To ensure a means of egress, the backup team may need to conduct frequent recons back to the door. The backup crew may also need to pull suspended ceilings to check for fire. If backup members do find considerable fire in the plenum, should they begin to darken it down? The answer is a qualified yes, for two reasons. First, they shouldn't push fire onto the attack or search crews ahead of their position. Second, it isn't the purpose of a backup team to be diverted into an attack. If these members happen to find a small, isolated fire in the plenum, they can knock it down, then continue their mission of monitoring the safety of the other crews. If the fire in the plenum is of some magnitude, they should notify the IC, who will then make the decision as to how to deal with it.

The members of a rapid intervention team will have one major focus at a fire in a professional building: collapse. These are lightweight buildings, highly prone to the effects of fire. The presence of interior partitions might make such an event survivable, but even this is no guarantee. Although large, these buildings aren't huge, and a 360 walkaround will prove useful. In outer design, professional buildings are usually very symmetrical or highly erratic. A walkaround will provide clues as to what the structure offers, as well as its hazards. Because of their lightweight composition and often problematic means of access, I liken these structures to garden apartment buildings. Many of them are set back from the parking lot and street, so aerials may not be able to reach them. In other cases, however, the parking lot may completely surround the building, and doors may be abundant on all sides. Naturally, the RIT should stage at a location that provides the best access to the crews inside. If the fire is on Side A, then hopefully the attack started from Side C. That's where the RIT should stage.

You'll generally only get one chance at putting out a fire in a professional building. The inherent potential for high levels of heat will force you to work quickly. Pull big lines and get them inside as soon as possible. The high-heat potential also makes for exposure problems. Some of these buildings are spaced closely enough that radiant heat will become a factor. Standard protection is in order, with the exception that aerial devices, rather than handlines, should probably be used to protect the most threatened exposures. Remember that glass only stops about 15 percent of radiated heat. If a nearby exposure has panel or curtain glass walls, it's imperative that a crew get inside to check conditions. One issue that often arises in such a situation is whether to force the doors or wait for a responsible party with keys. If there is any doubt, force the door.

Scenario

On the south end of the city is a relatively new development with five restaurants, three motels, and about a dozen new office buildings. Public streets with curbs and city hydrants on eight-inch mains run throughout the development. Some of the newer office buildings have gone up on spec, meaning that they have been erected by the developer, who hopes that he will be able to lease them by the time the buildings have been completed. Some of these lie vacant, and others are only partially occupied.

One of these buildings, a one-story wood frame, is about three-fourths occupied. It measures 100 X 150 feet and sits atop a concrete slab. The exterior load-bearing walls have 2- X 4-inch studs, 16 inches on center. The interior walls are of a similar construction and finished with plasterboard. Some of the walls have been papered, and others have a thin coat of plaster and paint. The floor is carpeted. On top of the structure is a truss hip roof. The top of this massive truss assembly is 24 feet above the eaves.

There are six entrances to the building, and over each is a gabled dormer with a window at the rake end. These windows are decorative, and although they provide access to the truss area, they are of no benefit due to the truss assembly in the truss loft.

Each entrance leads to a hallway lined with offices. The suites on the ends are larger than the rest, taking up the entire side. Both of these are occupied by physicians. The units between them have been subdivided to suit each tenant, and there is no rhyme or reason as to their size. Most of the tenants are involved in the medical profession, but you'll also find a

The case study of this chapter deals with a fire in a multi-occupancy modern office building. *Credit Coleman*

The case study as seen from sides "A" and "B".

couple of financial planners and one attorney. On any given workday, there are about forty-five people at work inside, plus about sixty patients and clients.

The building has no built-in fire protection. There is a central-station fire and burglar alarm.

Small-Department Response

Given a moderate amount of fire and sufficient resources on the first alarm, the IC feels that he can mount an aggressive attack. He assigns the first two engines as a single attack unit. Their objective is to take two lines against the fire. The members off the truck follow behind the nozzle and perform search. If the fire has vented itself via the windows, outside ventilation won't be necessary. Hopefully, too, the flames haven't yet entered the truss loft. With a quick knockdown, they won't.

Already out of units, the IC calls for a second alarm. With a couple of engines and a truck, he barely has enough people to handle the fire. In a worst-case scenario, he'll need far more. He asks for a third alarm just to achieve some level of comfort.

With more help on the scene, the IC orders bigger lines inside, and he has crews start to concentrate on what's happening above them. They pull ceilings as they go along. If interior operations are to be maintained, it's imperative that the truss loft be controlled. Increasingly,

The case study as seen from sides "A" and "D".

time is a factor. Once it can be determined that no lives will be lost, then the building, which some may view as disposable, will seem a little less important. After the all-clears have been sounded, the truss loft becomes the focus. If the roof vents itself, there will be a natural flow upward and outward. The crews will be able to move back, pulls ceilings, and flow water into the truss loft to minimize spread. Such a fire can only be controlled from below. To throw water at it from the outside wouldn't control the flames in the loft. If you want to save the building, you must fight the fire from inside. Knowing when and when not to attempt this comes with experience.

As always, a RIT should be assigned as soon as possible, as either a primary or a secondary assignment. After knockdown, ventilation, and search have been completed, overhaul and salvage can begin. In this sort of building, given its lightweight construction and chance of collapse, overhaul must be done cautiously. Set up collapse zones, and let a knowledgeable person perform a recon inside before commencing further interior operations.

Medium-size and Large-Department Response

Aggressive and properly placed attack lines will spell the difference between success and failure at this fire. The initial efforts should be aimed at cutting off the spread from uninvolved areas and rapidly knocking down the flames. Put out the fire, and most of your other troubles will go away. Once the attack crews are in place and advancing, send in backup crews behind them. If the fire vents itself by eating through the roof, so much the better. If the interior crews are fatigued, bring them out and flow water from an aerial for a minute or so, then put second-alarm crews inside to finish off the fire. Conduct overhaul and salvage per the usual advisories.

Chapter Nineteen Questions

1. Most modern low rise professional building are ____________ construction.
2. Architects must consider and plan for fires in new low-rise professional buildings. True or False
3. Why are most of the modern low-rise professional buildings two stories or less?
4. Most of the roof assemblies in these occupancies are ______________.
5. Most of the floor assemblies in these occupancies are ______________
6. The ___________ is when fire wraps around the exterior of an open center core office building.
7. Topside ventilation can be done any time fire enters the truss loft. True or False
8. To ensure a means of egress, backup must make several _______ back to the door of entry.
9. Fire can extend in the area between the ceiling and the floor above called the __________.
10. The members of RIT have as their general focus ______________.

Questions for discussion

a) Discuss the difference between a curtain and a panel wall.
b) Discuss the effects of PPV on a truss loft in a modern office building.
c) Discuss the positive effects of the fire venting itself through the roof of a truss roof assembly as opposed to the fire not venting itself with heavy fire in the same area.

Answers

Chapter 1

1. role
2. NFPA 1561
3. Focus
4. 1 to 5
5. Operations, Planning, Logistics, Administration
6. Core
7. The command post
8. Logistics
9. Safety, Liaison, Information Officer
10. Liaison

Chapter 2

1. Sector
2. Functional and Geographic
3. Functional
4. Geographic
5. To fine-tune expectations

Chapter 3

1. Command
2. Normal, Extreme
3. Know what you need and where to get it.
4. Yes
5. Give you a picture of what you'll need before the incident.

Chapter 4

1. We will take great risk to save life – We will take some risk to save property – We will risk nothing for life or property already lost.
2. Whether we can enter and operate effectively inside.
3. Already lost
4. Training officers to make educated, realistic decisions about expected outcomes.
5. Change - Conditions change themselves.

Chapter 5

1. Tradition, Firefighters are proactive, and Our reluctance to speak over the radio
2. Track the location of firefighters inside and outside.
3. PAR
4. Routine and Emergency
5. 15 minutes

Chapter 6

1. What to expect of his crews, What can and can't be done and What to expect from himself
2. Mutiny
3. 4
4. Non-essential crews
5. Send them back to the station for a defusing and then remain at the station for the remainder of the tour.

Chapter 7

1. The Dominant Vertical Channel or the interior stairs
2. False
3. Header courses
4. Cockloft
5. Joist
6. The interior stairway
7. 1 3/4"
8. False
9. Closest to the fire where savable victims could be located.
10. The same route as the attack line.
11. True
12. False
13. At the Command Post
14. RIT officer
15. The "what-ifs"

Chapter 8

1. 4
2. Access due to parking
3. Platform
4. Trusses
5. Mirror
6. One ladder and two firefighters in full turnouts
7. 1 3/4"
8. Oriented method
9. The apartment on the same side of the fire apartment, then the two apartments on the other side of the hall.
10. Over the hallway
11. Stairway
12. Search
13. Moisture
14. Attic ladder
15. 4

Chapter 9

1. Strip Mall
2. Early collapse
3. False
4. False
5. False
6. Concrete slab
7. Lumberyard
8. Check for extension
9. LxW÷3
10. That the roof has vented itself upon arrival.
11. Follow the attack line in.
12. Find the interior attack crew and lessen opposing lines.
13. At the upper levels of the party wall
14. Early collapse potential
15. Hydraulic

Chapter 10

1. True
2. Firefighter safety
3. False
4. Wide open or tight
5. 10 to 20 minutes
6. Unburned portion
7. Indirect attack off of the ceiling
8. 2 1/2" line with a smooth bore nozzle
9. Elevated streams from master appliances
10. Oriented method searching off of a 2 1/2" line
11. 10 x 10
12. Skylights
13. Get inside the exposure and check for extension.
14. Construction features
15. One inside and one outside

Chapter 11

1. Confining
2. Rescue
3. Defend in place
4. 1 victim for every 2 rescuers
5. Aggressive attack
6. The placement of the first line
7. Attack crew
8. After the flames have been knocked down and the smoke is cleared.
9. Triage
10. Aggressive ventilation

Chapter 12

1. Conflagration
2. Convection and radiation
3. False
4. False
5. Tripping and fatigue
6. Confine the fire to the smallest area possible
7. 3 times
8. Flank the fire
9. The direction of the attack and the amount of water needed.
10. 2 1/2" – Smooth bore nozzles
11. 2 1/2" line
12. At the ceiling from below
13. 800 degrees for 5 minutes
14. False
15. False

Chapter 13

1. 50,000 to more than 100,000
2. False
3. Ordinary
4. The time of day and the season
5. True
6. Life safety
7. True
8. False
9. True
10. Staffing and time
11. False
12. The size and location of the fire
13. False
14. One hour
15. Operations

Chapter 14

1. Life safety and rapid fire spread
2. False
3. Preplanning
4. Put the fire out!
5. So as to confine the fire to the smallest area possible.
6. The size of the air cylinder on his back.
7. False
8. Rescue Group
9. The area between the backup line and the attack or search crew and the area back to the exit.
10. Egress and the "what-ifs" of roof collapse.

Chapter 15

1. True
2. Hydraulic overhaul
3. Fixed
4. Movable fold-up chairs
5. Isn't
6. 10 to 20 minutes
7. 2 1/2"
8. Pews
9. True
10. Working off of a 2 1/2" line

Chapter 16

1. Life safety
2. False
3. Manpower
4. One rescue group per victim per every two to five minutes
5. Get a line between the majority of the residents and the fire.
6. True
7. True
8. Lack of manpower
9. Following the attack line in.
10. Get a line between RIT and the fire.

Chapter 17

1. The fire and the hazardous materials
2. False
3. True
4. The safety of firefighters and civilians
5. No fire, only heavy smoke is visible.
6. True
7. 2 1/2" line
8. Two members can split off and check the progress of the attack crew and also check the way out.
9. A little closer
10. Backup

Chapter 18

1. Life safety - Panic
2. The workers at the front desk
3. Platform
4. True
5. Venturi
6. 3 crews
7. To maintain a crew on the lane at all times
8. The walls in the hallway
9. True
10. On the stairway

Chapter 19

1. Truss
2. False
3. They don't require sprinklers.
4. Truss
5. Concrete slab
6. Doughnut effect
7. False
8. Recons
9. Plenum
10. Collapse

Index

A

Accessibility, 79-80, 87, 111, 122, 132, 134, 158, 169, 190, 192, 213, 215, 231-233, 248-250

Accountability (at incidents), 47-58: standard, 52; procedures, 52; system participation, 52, 55; accountability system, 54-58; levels, 56-57; questions, 58

Accountability board/sheet, 49, 55-57

Accountability system, 54-58: purpose, 54; responsibility, 55-56; crew riding list/integrity, 56; levels of accountability, 56-57; personnel accountability reports, 57; emergency evacuation, 58

Administration section, 19-20

Answers (chapter questions), 305-312

Apartments (garden), 87-110: scenario, 104-108; small-department response, 105-107; medium-size department response, 107-108; large-department response, 108; questions, 109

Apartments (older buildings), 67-86: scenario, 80-85; small-department response, 81-83; medium-size department response, 83-84; large-department response, 84-85; questions, 86

Attack line, 72-74, 81, 83-84, 96-97, 100, 115, 117, 121, 125-126, 133, 135-137, 140, 147, 149, 158, 164-166, 173-176, 179, 181-185, 191-192, 194, 200, 202, 208-211, 229-230, 247, 264, 279-280, 295

Attics, 71

B

Backup line, 73-74, 77, 100-101, 120-121, 137, 140, 160, 178, 196-197, 233, 249, 267, 297-298

Balconies, 94-95

Balloon construction, 67, 69-71, 273

Breaching walls, 62

Building types (fireground), 65-303: older apartments, 67-86; garden apartments, 87-110; strip malls, 111-128; vacant commercial occupancies, 129-150; institutional occupancies, 151-167; lumberyards, 169-186; enclosed malls, 187-203; restaurants, 205-222; churches, 223-238; nursing homes, 239-257; industrial fires, 259-272; hotels, 273-289; professional offices, 291-202

Building/structure access, 79-80, 87, 111, 122, 132, 134, 158, 169, 190, 192, 213, 215, 231-233, 248-250

Business/office buildings, 291-303. SEE ALSO Vacant buildings.

C

Checks and balances, 5-6

Churches, 223-238: scenario, 234-237; small-department response, 235-237; medium-size department response, 237; large-department response, 237; questions, 238

Claims unit, 19-20

Code assignment, 135, 138-139

Code standardization, 37-38

Collapse hazard, 93-94, 111, 115, 122, 131-132, 135, 172, 227, 229

Combination sectors, 23, 26-30

Command board/post, 6, 12-14, 17, 28-29

Command team, 6, 12-16

Commercial occupancies (vacant), 129-150

Communication, 14-15, 18-19, 21, 55-56, 63-64, 145

Communication span, 14-15, 21

Communications unit, 18-19

Company management, 52-53, 56

Compliance with standards, 54

Control span, 23

Corbel structure, 70

Cost unit, 20

Crew management, 52-53, 56

Crew riding list/integrity, 56

D

Decision-making rules, 41-45

Distress call, 59, 61

Division location, 25, 30

E

Emergency evacuation. SEE Rescue operation.

Emergency services, 50-51, 198, 250

Enclosed malls, 187-203: scenario, 198-202; small-department response, 200-202; medium-size department response, 202; large-department response, 202; questions, 203. SEE ALSO Strip malls.

Equipment availability, 34-37

Exposure, 25-26, 30, 115, 123, 130, 133, 144-145, 227

Extension check, 78-79, 121-123, 143-144

Extrication operation, 26

F

Facilities unit, 18-19

Fatal fire, 38

Fire flow formula, 117

Fire load, 88, 121, 172, 175, 291, 294

Fire protection, 44

Fire spreading, 67-80, 90-91, 117, 135, 156, 243-244, 276, 293, 295

Fireground, 65-303: apartment buildings (older), 67-86; garden apartments, 87-110; strip malls, 111-128; vacant commercial occupancies, 129-150; institutional occupancies, 151-167; lumberyards, 169-186; enclosed malls, 187-203; restaurants, 205-222; churches, 223-238; nursing homes, 239-257; industrial fires, 259-272; hotels, 273-289; professional buildings, 291-202

Floor assembly, 70-71, 90, 115, 155-156, 171-172, 189, 207, 241-242, 275, 293

Food service buildings, 205-222: scenario, 218-221; small-department response, 219; medium-size department response, 219-221; large-department response, 219-221; questions, 222

Food unit, 18-19

Forcible entry, 81, 83-84, 95, 108, 115-116, 128

Functional sectors, 23-30

G

Garden apartments. SEE Apartments (garden).

Geographic positioning system, 47

Geographic sectors, 23-30

Ground support unit, 18-19

H

Hazardous materials, 13, 259-260, 262, 268

Hospitals, 151-152, 156, 160. SEE ALSO Institutional occupancies.

Hotels, 273-289: venting low rise, 284-284; scenario, 284-288; small-department response, 286-287; medium-size department response, 288; large-department response, 288; questions, 289

Housing/living quarters, 67-110, 154, 158, 160, 239-257: apartments (older buildings), 67-86; apartments (garden), 87-110; nursing homes, 154, 158, 160, 239-257. SEE ALSO Institutional occupancies.

I

Incident commander, 11-12, 17, et passim

Incident management system, 3-22: questions, 22

Incident-specific assignment, 51, 54

Incident-specific operations, 4

Incident-specific resources, 34

Industrial fires, 259-272: scenario, 268-271; small-department response, 269-271; medium-size department response, 271; large-department response, 271; questions, 272

Information officer, 20-21

Institutional occupancies, 151-167: scenario, 160-166; small-department response, 163-165; medium-size department response, 165-166; large-department response, 166; questions, 167. SEE ALSO Nursing homes.

L

Large-department response, 84-85, 108, 127-128, 149, 166, 185, 202, 219-221, 237, 255-256, 271, 288, 302

Liaison officer, 20

Life hazard, 42, 59-64, 71, 83, 88, 90, 92-98, 101, 106-107, 111-113, 126-132, 135, 145, 151, 153, 155-156, 158, 160, 163-164, 169, 173, 187-188, 191, 205, 207, 216, 223, 227, 239-257, 259-260, 262, 267-268, 273, 276

Logistics section, 18-19

Lumberyards, 169-186: scenario, 179-185; small-department response, 179-183; medium-size department response, 183-184; large-department response, 185; questions, 186

M

Mall fires, 111-128, 187-203. SEE ALSO Enclosed malls AND Strip malls.

Mayday management, 59-64: questions, 64

Medical/rehabilitation unit, 18-19

Medium-size department response, 83-84, 107-108, 126-127, 148-149, 165-166, 183-184, 202, 219-221, 237, 254-255, 271, 288, 302

Missing firefighter, 59-64

Multitasking, 63-64

Mutiny (personnel), 59-60

Mutual-aid agreement, 35-37, 54

N

National Fire Academy, 3-4, 23, 41-42: risk policy, 41-42

National Fire Protection Association, 7-8, 47, 50-54, 117

NFPA Standard 1500, 47, 52-53

NFPA Standard 1561, 7-8, 47, 50-51

Nursing homes, 154, 158, 160, 239-257: scenario, 250-256; small-department response, 252-254; medium-size department response, 254-255; large-department response, 255-256; questions, 257. SEE ALSO Institutional occupancies.

O

Occupational safety and health program, 52-53

Occupied buildings. SEE Housing/living quarters AND Hotels AND Institutional occupations.

Office buildings, 291-303: scenario, 299-302; small-department response, 301-302; medium-size department response, 302; large-department response, 302; questions, 303

Old-law tenements, 80-85

On-scene crew (mayday), 60

Operations officer, 15-17

Outside ventilation man, 73, 83-84

Overhauling, 84-85, 101, 107-108, 124, 126, 149, 165, 173, 182-183, 202, 237

P

Passport system, 50, 53

Penal institutions, 151, 153, 155-157, 160. SEE ALSO Institutional occupancies.

Personnel, 5, 36, 44, 47-64, 81, 83: training program, 44, 59-64; control, 47-60; tracking, 47-58; accountability system, 54-58

Personnel accountability, 5, 47-61, 64: standard, 52; procedures, 52; system participation, 52, 55; accountability system, 54-58; levels, 56-57; questions, 58

Personnel accountability report, 5, 50, 54, 56-58, 61, 64

Personnel control, 47-60: mutiny, 59-60

Personnel identification, 51

Personnel tracking, 47-58

Petroleum refinery, 35

Planning section, 17-18

Plants (industrial), 259-272: scenario, 268-271; small-department response, 269-271; medium-size department response, 271; large-department response, 271; questions, 272

Platform construction, 69, 89

Positive-pressure ventilation, 62-63, 75, 98, 107-108, 120, 122, 212, 215, 229, 245, 266, 283, 296

Pre-incident assignment, 4, 54

Pre-incident planning, 69-80

Prisons. SEE Penal institutions.

Procurement unit, 19-20

Professional buildings, 291-303: scenario, 299-302; small-department response, 301-302; medium-size department response, 302; large-department response, 302; questions, 303

Property risk, 42-43

Q

Questions, 22, 31, 39, 45, 58, 64, 86, 109, 128, 150, 167, 186, 203, 222, 238, 257, 272, 289, 303

R

Radios, 37, 48

Railroad flats, 80-85

Rapid intervention team, 5, 59-62, 79-80, 83, 85, 102-104, 145, 178-179, 184-185, 197-198, 216-218, 234, 249, 267, 284, 298-299

Rescue operation, 26, 51, 58, 83-85, 92-93, 95-98, 101, 106-107, 126-128, 141, 148, 151, 153, 194-198, 205, 215-216, 244-250, 266, 280-281, 283, 295

Resource allocation, 33-39, 44: questions, 39

Resource availability, 34-37

Response (form of), 1-64: incident management system, 3-22; sectoring large incidents, 23-31; resource allocation, 33-39; rules of engagement, 41-45; risk assessment, 41-45; accountability at incidents, 47-58; mayday management, 59-64

Response time, 44

Responsibility (accountability), 55-56

Restaurants, 205-222: scenario, 218-221; small-department response, 219; medium-size department response, 219-221; large-department response, 219-221; questions, 222

Risk/risk assessment, 41-45, 129-130, 132, 135: rules of engagement, 41-45; questions, 45

Roof assembly, 69-70, 89, 91, 114-115, 122, 124, 137, 142-143, 153, 155, 171, 189, 207, 223, 225, 241-243, 262, 275, 294, 296

Roster system, 49

Rules of engagement, 41-45: risk assessment, 41-45; questions, 45

Rural departments, 11, 17

Rural fire, 34, 51

S

Safety, 20, 42, 52-53, 59-64, 71, 83, 88, 90, 92-98, 101, 106-107, 111-113, 126-132, 135, 145, 151, 153, 155-156, 158, 160, 163-164, 169, 173, 187-188, 191, 205, 207, 216, 223, 227, 239-257, 259-260, 262, 267-268, 273, 276

Safety officer, 20, 145

Salvaging, 83-85, 107-108, 126, 198, 233-234, 237

SARA III mandates, 8
Scenarios (fireground), 80-85, 104-108, 124-128, 145-149, 160-166, 179-185, 198-202, 218-221, 234-237, 250-256, 268-271, 284-288, 299-302
Schools, 151, 153-154, 156, 159-166. SEE ALSO Institutional occupancies.
Search operation, 26, 77, 83-85, 92-93, 95-98, 101, 106-107, 126-128, 140-141, 158, 160, 177, 182, 194-198, 205, 210-213, 215-216, 231, 233, 244-250, 266, 280-281, 283, 295
Sectoring large incidents, 23-31: questions, 31
Ship facilities, 36
Shopping malls, 111-128, 187-203: strip malls, 111-128; enclosed malls, 187-203
Situation status unit, 18
Small-department response, 81-83, 105-107, 125-126, 147-148, 163-165, 179-183, 200-202, 219, 235-237, 252-254, 269-271, 286-287, 301-302
Staffing constraints, 36, 61, 81, 83
Staffing level, 36
Stairways, 67, 71, 73-74, 80-81, 85
Standard operating guidelines, 52-53
Standardization, 5, 7-9, 37-38
Strategies/tactics, 4, 63-64
Strip malls, 111-128: scenario, 124-128; small-department response, 125-126; medium-size department response, 126-127; large-department response, 127-128; questions, 128. SEE ALSO Enclosed malls.
Structure analysis, 67-81, 87-105, 111-125, 130, 132, 135, 145, 147, 153, 155-156, 160, 162-163, 169-172, 187-191, 207, 223, 225, 227, 239-245, 260, 262, 267, 273, 275-276, 279, 281, 283, 291, 293-299
Subdivided structures, 67-86
Supervisory responsibility, 52-53
Supply unit, 18-19
Support branch, 18-19

T

Task assignment, 4-7, 10, 12-16, 23-30
Teamwork, 50
Terminology, 37-38
Thermal imaging, 264
Think tank strategy, 63-64
Time unit, 19-20
Training program, 44, 59-64: decision-making, 44
Transferring command, 8
Turnout gear, 60-61

V

Vacant buildings, 43, 129-150
Vacant commercial occupancies, 129-150: scenario, 145-149; small-department
response, 147-148; medium-size department response, 148-149; large-department
response, 149; questions, 150
Ventilation, 62-63, 73-75, 93, 98, 106-108, 118, 120, 122, 141-143, 153, 158, 177, 191, 212, 215, 217, 223, 225, 229-230, 232, 245-247, 264-266, 276, 283-284, 296-297
Ventilation crew, 73-75
Venting low rise, 283-284

W

Wall assembly, 62, 67, 69, 78-79, 91-92, 102, 113-114, 153, 171, 188-189, 228, 241-243, 273, 276, 291, 293
Wildland fire, 7
Wood construction, 88-89, 102